SQC/SPC

MANUFACTURING EXPERIENCES

Thomas J. Drozda
Editor

Robert E. King
Manager

Joy Beall
Production Assistant

Published by

Society of Manufacturing Engineers
Publications Development Department
Reference Publications Division
One SME Drive, P.O. Box 930
Dearborn, Michigan 48121

SQC/SPC
MANUFACTURING EXPERIENCES

Library of Congress Catalog Card Number: 89-61109
International Standard Book Number: 0-87263-362-4
Manufactured in the United States of America

SME wishes to express its acknowledgement and appreciation to the following contributors for supplying the various articles reprinted within the contents of this book.
Appreciation is also extended to the authors of papers presented
at SME conferences or programs as well as to the
authors who generously allowed publication
of their private work.

Fabricators and Manufacturers Association International
5411 East State Street
Rockford, Illinois 61108

Journal of Quality Control
American Society for Quality Control
310 West Wisconsin Avenue
Milwaukee, Wisconsin 53203

Manufacturing Engineering
Society of Manufacturing Engineers
One SME Drive
P.O. Box 930
Dearborn, Michigan 48121

Modern Casting
American Foundrymen's Society, Incorporated
Golf and Wolf RDS
Des Plaines, Illinois 60016

Society of Manufacturing Engineers
One SME Drive
P.O. Box 930
Dearborn, Michigan 48121

Tooling and Production
6521 Davis Industrial Parkway
Solon, Ohio 44139

PREFACE

Today, the quality of products and services are more in focus than ever as every industrial country continues the struggle to maintain or, better yet, improve its competitive position in the global arena.

Consequently, manufacturing engineers and managers are keenly aware that they must maintain and/or improve the quality of the goods and services they offer for sale in both domestic and international markets. And they have received that message from their most significant and influential source of input — their customers.

To truly satisfy the needs and demands of the customer, a product must adequately fulfill the purpose for which it is intended. Additional concerns include product life, dependability, and safety, ease of maintenance, the availability of spare parts and service facilities, and clear instructions for proper product use.

How do we meet these demands and, thus, achieve our goals? This book is designed to offer considerable insight into answering that very question.

This book attempts to provide overviews in the areas of both statistical process control (SPC) and statistical quality control (SQC). The papers presented in this volume outline some successful applications in the quality area and, it is hoped, will provide valuable input in establishing quality control programs. The ways in which quality assurance practices differ from one company to another will become clear. In some industries, they may consist entirely of inspection-related activities, whereas, in other areas they may include prevention and inspection in varying proportions.

I would like to thank the companies, organizations, publishers, and authors whose material appears in this book. Appreciation must also be extended to our Publications Development Department staff for their assistance in the research and development required to make this book possible.

Thomas J. Drozda
Director of Publications
Society of Manufacturing Engineers

ABOUT THE
EDITOR

Tom Drozda is the Director of the Publications Division at the Society of Manufacturing Engineers. Mr. Drozda has served as Editor of the Society's *Tool and Manufacturing Engineers Handbook* and as Editor-In-Chief of the Society's *Manufacturing Engineering* magazine.

 Mr Drozda's background includes process engineering experience at Detroit Diesel Allison Division of General Motors. He held the position of engineering editor at *Production* magazine where he developed many articles exploring a wide range of metalworking subjects including tolerance control and CAD/CAM technology.

Although an editor by trade, Mr. Drozda holds a professional engineering license in the state of Michigan and is a Certified Manufacturing Engineer. Mr. Drozda holds a Bachelor of Science in Industrial Engineering and a Master's degree in Business Administration.

SME

The informative volumes of the Manufacturing Update Series are part of the Society of Manufacturing Engineers' many faceted efforts to provide the latest information and developments in engineering.

Technology is constantly evolving. To be successful, today's engineers must keep pace with the torrent of information that appears each day. To meet this need, SME provides, in addition to the Manufacturing Update Series, many opportunities in continuing education for its members.

These opportunities include:

• Monthly meetings through five associations and their more than 300 chapters and 165 student chapters worldwide to provide a forum for membership participation and involvement.

• Educational programs including seminars, clinics, programmed learning courses, as well as videotapes and films.

• Conferences and expositions which enable engineers and managers to examine the latest manufacturing concepts and technology.

• Information on Technology in Manufacturing Engineering database containing technical papers and publication articles in abstracted form. Other databases are also accessible through SME.

The SME Manufacturing Engineering Certification Institute formally recognizes manufacturing engineers and technologists for their technical expertise and knowledge acquired through experience and education.

The Manufacturing Engineering Education Foundation was created by SME to improve productivity through education. The foundation provides financial support for equipment development, laboratory instruction, fellowships, library expansion, and research.

SME is an international technical society dedicated to advancing scientific knowledge in the field of manufacturing. SME has more than 80,000 members in 70 countries and serves as a forum for engineers and managers to share ideas, information, and accomplishments.

The society works continuously with organizations such as the American National Standards Institute, the International Organization for Standardization, and others, to establish and maintain the highest professional standards.

As a leader among professional societies, SME assesses industry trends, then interprets and disseminates the information. SME members have discovered that their membership broadens their knowledge and experience throughout their careers. The Society of Manufacturing Engineers is truly industry's partner in productivity.

MANUFACTURING UPDATE SERIES

Published by the Society of Manufacturing Engineers and its affiliated societies, the Manufacturing Update Series provides significant up-to-date information on a variety of topics relating to manufacturing. This series is intended for engineers working in the field, technical and research libraries, and as reference material for educational institutions.

The information contained in this volume doesn't stop at merely providing the basic data to solve practical shop problems. It also can provide the fundamental concepts for engineers who are reviewing a subject for the first time to discover the state of the art before undertaking new research or applications. Each volume of this series is a gathering of journal articles, technical papers, and reports that have been reprinted with expressed permission from the various authors, publishers, or companies identified within the book. Educators, engineers, and managers working within industry are responsible for the selection of material in this series.

We sincerely hope that the information collected in this publication will be of value to you and your company. If you feel there is a shortage of technical information on a specific manufacturing area, please let us know. Send your thoughts to the Manager, Publications Development, Reference Publications Division at SME. Your request will be considered for possible publication by SME or its affiliated societies.

TABLE OF CONTENTS

CHAPTERS

CHAPTER 1

INTRODUCTION

Reprinted from *Manufacturing Engineering*, August 1987

Developing a Strategy for Quality

By John M. Martin

Quality appears, at least temporarily, to be vying with CIM (computer-integrated manufacturing) as the topic on the tip of everyone's tongue in manufacturing. Of course, the two subjects are not opposites, but rather they are both key elements in the drive to recapture US preeminence in the manufacturing arena.

Quality is simply the result of paying attention to detail and doing things the way they should be done. We, as a manufacturing nation, by virtue of the drubbing we have taken in the world marketplace, have finally and begrudgingly accepted the fact that has been staring us in the face these past 15 years.

Namely, that we have to improve the quality of our products, indeed, the entire way we do business, in order to

be ranked as contenders in those areas where our competitiveness has flagged. A quality product, competitively priced, delivered on time, and properly serviced throughout its life cycle, is the only true strategy for regaining that competitiveness.

Developing a strategy for quality is not just a laudable goal. It is a method of survival. It must begin at the top, be communicated throughout the company, and be organized for constant feedback, reinforcement, and evolution.

The results will be increased customer satisfaction, reduced costs, the potential for greater sales, revenue, and market share, and a more productive and hassle-free working environment on both the shop floor and in the office.

Developing a strategy for quality must begin at the highest levels of the

corporation. That's where it started at Lockheed-Georgia Co. (Marietta, GA), a division of Lockheed Corp., where President Ken Cannestra called in a team of key management in November 1986 to start the wheel turning in what would evolve into that corporation's Total Quality Improvement Program (TQIP). At IBM (Armonk, NY), Vice President for Quality William W. Eggleston reports that it was former Chair Frank T. Cary who, in the late 1970s, significantly ramped up the emphasis on quality in a company already well known for its substantial quality commitment.

Top management's stand on quality must be more than words and good intentions, it must be "policy and resources, plans, budgets, and systems," says Dr. Armand V. Feigenbaum, respected quality proponent and president of General Systems, Inc. (Pittsfield, MA), which designs and installs total quality operational and control systems for quality control implementations. Adds Kendell Johnson, president of Robosoft Corp. (Apple Valley, MN), which is designing and marketing a universal controller and programming system for robots, "At the very outset, top management must consciously decide if it is really personally dedicated to quality. Are you ready to hold back shipments and jeopardize weekly and monthly revenue goals to fulfill the mandates of a true quality program?"

When that CEO or top management commitment is lacking, why is it lacking? "There's a lot of tugging at the

IBM employees at a reliability analysis class learn how to interpret product test data at its training school in Thornwood, NY.

John M. Martin is a New York-based writer and consultant in manufacturing and computer-integrated manufacturing.

boss' shirt," says Eggleston, who stresses that quality "must still be the priority." Johnson adds the thought that "Manufacturing has been a second-class citizen in this country for 25 years. Most current top management took other channels to the top, like marketing and finance, and the quality issue in manufacturing is not intuitively obvious to them because of their background."

Others decry the short-term outlook of a management said to be obsessed with quarterly performance. While there is certainly some truth to that criticism, it could also be added that it's not very easy for management to plan "in a governmentally influenced environment in which things like interest rates, investment tax credits, depreciation schedules, and the value of the dollar all gyrate wildly and vary from administration to administration," says a source at the US Department of Commerce.

In any event, the reasons why are less important than that things get changed. And when top management does believe and buy into the quality issue, it must also realize that quality is no magic wand, but rather is something that requires "time, diligence, and investment," says Eggleston. "You will get some quick returns, but the real payoff for the company's competitiveness lies in the long-term implementation." He adds: "Of course, this can be difficult for a CEO—very few of them got their job because of patience. They are typically doers, drivers, people who react swiftly and go after things."

T hose very attributes, however, properly harnessed, can be the key to success for a quality program. Because once the commitment has been made at the top, a tone must be established for the rest of the company. It must become clear to everyone that quality is more than lip service, the latest "fad-of-the-week."

"One problem with quality is that the word is like 'motherhood' or 'patriotism'," states George H. Kuper, executive director of the Manufacturing Studies Board at the National Research Council (Washington, DC). "But actions transmit real intent, and you'll often see a bunch of quality posters and charts on the wall while the foreman tells the line workers to disregard the quality issue."

One way to set that tone is to communicate the quality message throughout the corporation and then

Lockheed-Georgia (Marietta, GA) uses a patented anticorrosion process to ensure better products for its customers.

set up the organizational structures and procedures to bring it into being and make it stick. Nothing can probably do that better than tying quality to performance reviews and career advancement.

At Giddings & Lewis Machine Tool Co. (Fond du Lac, WI), a division of AMCA International, Vice President and General Manager Orville W. Ehrhardt reports, "We have formalized performance evaluations for all employees, and we are beginning to build quality measurements into all of these. Each manager will have targets to demonstrate improvements in the quality area just like in the areas of cost containment and productivity."

The Computer Systems Div. of Harris Corp.(Fort Lauderdale, FL) has

its own way of stressing the relationship between quality and accountability. Director of Quality Dave Robinson talks about "ownership," where, for example, "Design and development people will share 'ownership' of the product not only during conception and concept review, but also through field performance and the entire product life cycle." At Giddings & Lewis, too, rates for time and cost associated with unsatisfactory quality are charged out to the departments responsible.

Interestingly, putting people on the spot on the quality issue also leads to examination of how interrelated the various functions of a corporation really are. Honing in on a quality problem in the manufacturing area may lead to a design specification that is not manufacturable, which may in turn lead back to a marketing directive for an enhanced performance characteristic that is destabilizing the product's reliability.

In fact, the very act of communicating the strategy for quality that is being developed within a manufacturing corporation is the key step toward organizing the firm's various departments and functionality into the kind of "open-channel" structure that is critical to both the success of a quality operation and the kind of cultural change that will make it endure.

There are a variety of ways to fashion the organizational structure for implementing the quality stra-

tegy. Each will be partially generic and partially indigenous to the particularities of the organization involved. For example, drawing on his experience at Harris Corp., Robinson states, "There are probably four major areas that need to be integrated into quality planning: design processes, manufacturing processes, raw materials and subassemblies, and people."

In design, as mentioned earlier, the hand-off to manufacturing at Harris is accompanied by a continued relationship to the product. In manufacturing, Harris employs a just-in-time system so that smaller lot deliveries can be inspected in real time for identifying quality-related problems early in the manufacturing cycle. Harris' raw material and subassembly strategies

At IBM's Lexington, KY, operation, an automated inserter places components in an electronic circuit board assembly and test module.

Top management's stand on quality must be more than words and good intentions

for quality include "reducing our number of suppliers and establishing a close business relationship with those that remain. JIT forces you to do that," Robinson says. Finally, on the people side, Harris is in the sixth year of its version of quality circles, which it calls improvement circles, and which now encompass some 49% of the staff.

At Lockheed-Georgia, the quality impetus from Cannestra led to the establishment of an internal consulting team of four key management individuals. They were charged with a sole responsibility: the planning and successful initiation of the quality stratagem. Cannestra subsequently appointed a program director to steer the implementation, and these functions will be phased out "only when the quality improvement process becomes part of our corporate culture," says Bob Meadows, director of TQIP.

The team began its work by consulting with other companies who had implemented quality strategies, as well as with quality experts. Synthesizing outside ideas and a few of its own, the team laid out a plan for the total involvement of the company, representing each of its functional branches: engineering, production operations, materiels, operations support, product assurance, contract management, information services, finance, marketing, and so forth.

These functional branches were given basic guidelines to follow. "We asked them to analyze their operations, figure out who their internal 'customers' and 'suppliers' were, identify what products and services they provided to their customers, identify what were the major product attributes needed to satisfy the customer, and use these to develop their measurement systems. But each of them was charged with developing a plan tailored to their own area, as opposed to a top-down, highly structured approach," Meadows says.

At this point the team involved the four original members and 16 representatives from the different functional branches. A plan was developed and submitted to Cannestra in late December. He reviewed it, and in late January 1987, Lockheed-Georgia kicked off the program with a major speech by Cannestra at Building B-3 in Marietta. All told, some eight different major presentations were made in different company locations to encompass Lockheed-Georgia's 19,000 employees. Internal newspapers and closed-circuit video presentations were also used to get the message across.

A new program at Giddings & Lewis was also initiated at the top, when President Louis J. Lawrence held a meeting with top managers at the company last December 15 to commission the quality effort. That first meeting was followed by another for the middle echelon of management, attended by Lawrence to show that he meant business. Meetings were then held with line supervisors and the technical and professional staffs, and a training program consisting of 14 hours of classroom work was used to reach shop floor staff. Ehrhardt states, "All of our educational plans will be completed by the third quarter of 1987, we've got until the end of the year to establish our quality targets, and we'll enter 1988 with the program in place."

The most important thing about all these efforts is not the organizational detail particular to each implementation, but rather the driving force of top management commitment that made the talk about quality turn into a program implementation. Nevertheless, certain aspects do emerge in each of the programs as key determinants of their potential or already realized success.

Also critical is establishing very specific targets for quality improvement. Harris, for example, has a corporate five-year business plan that specifically states quality objectives. Each department within the Computer Systems Div. has an annual operating plan with specific objectives for improving quality within the various departments. These plans are reviewed quarterly to make sure the quality objectives are being met.

Training is also paramount. "'How

The quality effort at Giddings & Lewis (Fond du Lac, WI) emanates from top management and involves all levels of the operation.

do I do it?' is what your workforce really needs to know," states Eggleston of IBM. "You have to build courses to train personnel in concepts and techniques. Quality improvement is a constantly changing thing that you learn as you go along, and you must redesign the training to keep pace as you move ahead." IBM has a quality school in Thornwood, NY, where employees undergo direct training and are also trained as teachers who will return to their home plants to instruct others in the concepts and techniques of quality.

Most of the companies developing a quality strategy also stressed the key role played by supplier relations. "If you're sending critical parts to outside manufacturers, you risk the results of different part interpretations and varying equipment tolerances, and if there are multiple vendors involved, chances are the variations themselves will all be different," says Robosoft's Johnson. The key is careful selection and monitoring of vendors, a comprehensive program of vendor management that rewards vendor commitment to quality as much as it does commitment to price and delivery.

Companies with the more promising and successful programs also seem to have established a high-level position to oversee quality within the corporation, with job titles like vice president for quality, director of quality, or vice president of quality assurance. Roger Hartel, who has the latter title at Allen-Bradley's Industrial Controls

Group (Milwaukee), calls himself a "rabble-rouser." Eggleston at IBM reports that he oversees "a very small staff at the corporate level whose sole purpose is to communicate and aid the quality effort throughout the organization." Each of these individuals also reports to superiors at or near the very top of their respective company's operations.

Many people seem to think that, in this time of intense cost cutting to remain competitive, a quality program is just too expensive. Nothing could be further from the truth. Forget all the "quality doesn't cost, it pays" slogans; listen to some hard talk from the experts.

Bob Meadows reports that at Lockheed-Georgia, "We received a good suggestion from one of the employee involvement teams participating in our TQIP program that might save us up to $2.5 million a year." Adds Eggleston at IBM: "I don't have one example of a quality enhancement that hasn't saved us money in productivity and work-in-process improvements." Giddings & Lewis' Ehrhardt states, "Quality emphasis often places a poor third behind cost and schedule, whereas in reality quality controls a lot of cost and, when you have to do things over, schedule as well."

The cost of quality to an organization that doesn't adequately address the issue can be enormous, easily approaching a figure of 20-25% of sales when you calculate how much is spent for manuals and educational materials, maintenance, inspection and testing, reworking bad parts, scrap, service calls and warranty charges, product litigation, and more. "We know it is more than 20%," says the quality vice president for a *Fortune* 500 company, "and as a company you have to be mad enough about that to take the steps to drive it down."

Developing a stratagem for quality in manufacturing, in fact, is in itself part of the strategy of cost containment. Ehrhardt says, "Cost containment is a real problem today, with the pressures on pricing and squeeze on profits. There's only so much you can do along those lines with staff and engineering and manufacturing solutions. The potential for cost reductions in the quality area, however, is really large."

That's because the key factor that ensures quality—consistent and controlled operations from the executive suite down to the factory floor—is vir-

tually synonymous with efficiency and productivity. The effort of a quality program to turn a company upside-down in order to examine its operations and institute an agreed-upon and improved set of work processes can only result in a beneficial cost equation as well.

IBM certainly seems to think so, as it is one of those few companies frank enough to state that the cost of not paying attention to quality consumes some 20% of its revenue. Eggleston of IBM reports that the last three to four years of its quality program have reduced that 20% figure by 10%. He additionally notes that most companies that get serious about quality find that 20% is not an unusual number.

Developing a strategy for quality in manufacturing must not only be driven from the top. It must also make the customer the paramount consideration in deciding just exactly what constitutes quality. Not *Caveat Emptor*, but *Satisfaciatur Emptor*—Let The Buyer Be Satisfied.

"Quality is what the customer, not engineering managment, says it is," says Feigenbaum of General Systems. "We uncover a deep bias in some manufacturing companies. They think that they know more than their customers do." Adds Dennis A. Swyt, deputy director at the Center for Manufacturing Engineering, a section of the National Bureau of Standards (Gaithersburg, MD), "The bottom line is customer satisfaction. Number one, you understand your customer's needs, desires, and expectations, and then you organize and develop the technology to produce that."

Feigenbaum says, "There is a basic change in the way people buy today. Eight out of 10 people today make quality equal to or greater than price as their main consideration, whereas only three or four out of 10 felt that way a decade ago. This is as major a shift in US marketplace trends as any that has occurred since World War II."

Robinson of Harris points out that quality is not only a customer's perception, but that it is also a moving target. "Quality definitely changes with time," he says. "Customer requirements change as their needs change." Harris keeps on top of this by working closely with its customer support group on customer satisfaction surveys that track not only the quality of the product, but also the quality of the service, support, and sales effort that Harris provides to the

customer. "Every month these surveys are pulled together and summarized so we can see what we are doing right and what we are doing wrong," Robinson says. "For example, we'll get suggestions about which hardware or software features should be improved or added. We'll take them and evaluate them, make direct contact with the customer who filled out the survey, and then make development assignments for product enhancements. Our general manager chairs these monthly meetings, and vice-president-level staff and other key people are in attendance."

Swyt refers to an article in the September-October 1983 issue of the *Harvard Business Review*, "Quality on the Line," by David A. Garvin. The article makes a detailed comparison between US and Japanese air-conditioner manufacturers and touches on a lot of interesting points, but what Swyt stresses is that "the Japanese manufacturers profiled made customer satisfaction the key variable in their manufacturing control group. They used an internal surrogate, setting up a 'customer group' composed of company workers inside the plant who were authorized to sample products at all intermediate stages before final shipment and who gave final approval on product design, production, and shipment."

Interestingly, defining quality in terms of customer satisfaction brings the process of developing a strategy for quality far beyond the specific manufacturing function and into all areas of the company. It can be a way to dramatically improve white-collar productivity by concentrating on the quality angle.

Ehrhardt of Giddings & Lewis

points out that insufficient quality can strike anywhere. "Typically, the emphasis is on work spoilage in manufacturing," he says, "but if sales fails to include all the customer specifications on the order, we have to make changes down the line. Maybe we have to reprocess the order through engineering. If all the final contract language isn't properly fixed, maybe the customer doesn't accept the invoice. Dimensional errors and changes made on the manufacturing floor can occur if the specifications aren't clear. There can be errors in the parts lists passed down to the floor from the engineering group—and so on."

One of the examples Eggleston reports involves a quality drive in IBM's order-entry billing system. "The customer is interested in accuracy," he says, "and we were unhappy with this system about three or four years ago. We were billing from a process that was designed some 20 years ago when our business was almost 100% rentals. It was probably the best rental billing system in the world, but we revolve around purchases now, not rentals. Reducing the defects in that system has given us a three-times improvement to date in billing quality, and that figure will increase even more when our long-term, newly designed systems are implemented."

In a January 27, 1987, speech to kick off his company's Total Quality Improvement Program, Lockheed-Georgia's Cannestra said, "Quality is a matter of survival. Quality will determine our market share. It may be the determinant of whether we win new business in the future. How the customer perceives us will determine whether we win the next competition."

Developing a strategy for quality in the manufacturing environment requires, according to Feigenbaum, "A decision at the boardroom level to provide leadership, putting in place an organization-wide implementation of the actual work practices and disciplines of quality, and making quality a habit, as well as an upward moving and continually improving target."

Quality is "not a department, but a set of work processes," Feigenbaum adds. Says Garvin in the preface of his 1983 *Harvard Business Review* article: "Superior levels of performance come ...from sound management practices deliberately and systematically applied."

Quality means getting your processes under control. It is not simple,

An IBM robotic manufacturing system completes an operation on a subassembly and assures quality in completed modules.

but it is straightforward, a long, painful process of "tweaking the line, getting back to basics, rolling up your sleeves, and digging into things. Everybody wants a magic pill, but there are none," says Robosoft's Johnson.

It requires an immense cooperative effort. Cannestra again: "Each of us has a customer. Whether you are filling out forms, or drilling holes, or answering questions, you're doing it for someone else. You've got to think of quality and quality improvement as doing a better job individually so that when you pass it on to someone else, it fulfills his or her requirements. You must think of the next person in line as your customer. You must know what is required to get the job done, so you can do a better job in what you're doing to satisfy those needs."

In undertaking the act of developing and implementing a quality strategy, a company will move, as IBM did, "from defect management—the concept of customer requirements, setting measurements, and managing defects out of the work—to a realization that the best way to improve quality is to design work that doesn't produce defects to develop new and better work processes, tooling, and materials to support the new product. In this process a company must undertake a total review of its enterprise to understand the customer's needs, the company's operations, and how the competition does it elsewhere," Eggleston says.

The process will never end. While

one company is getting its defect rate down to so many parts per thousand, a competitor from somewhere will be reaching for rates based on parts per million. "Quality only goes one of two ways: It either gets better or it gets worse," says Allen-Bradley's Hartel.

Developing a strategy for quality is a way to stop the bleeding and rejuvenate a business. Don't wait until it's too late. Decide to develop a quality strategy, find out what the customer thinks is not important, focus on those elements that satisfy the customer, figure out where the company is now, and develop the strategy to get from where you are now to where the customer wants you to be.

The goals, according to Hartel, are to "satisfy the customer's need, put yourself in a position to be better than the competition, and put yourself on a 'slope' to keep you better than the competition." The three essential elements of a strategy for quality excellence, he says, are "knowledge of the customer's needs, a planning system to take the knowledge and systematically turn it into a 'deliverable,' and the management attitude to communicate to the organization the importance of quality."

Above all, concludes Feigenbaum, don't make the quality strategy "a bunch of July 4th speeches and fix-up projects, just one more crusade that everyone knows will eventually die and, hopefully, be buried without an autopsy." ∎

Presented at the SME/FMA FABTECH Conference, June 1986

Quality: Complete Corporate Driver

Howard B. Aaron
Q.E.D.

PROLOGUE

Management and Motivation

As discussed in earlier publications (references 1,2) to improve quality (and hence productivity) requires long-term focus and commitment; understanding what needs to be done and how to do it (by all involved); planning and control at every step in the product life cycle; and attention to the details of execution. This requires more than merely trying to design "it" in; build "it" in; etc. It requires, above all else, that we manage it in. People will endeavor to do a good job provided they understand the requirements; know in what way(s) what they are doing deviates from expection; and have the knowledge and resources to get from where they are to where they need to be.

The management system must be formal, well managed, and contain "side-to-side" as well as "top-down" and "bottom-up" elements. Short-term goals must support and be consistent with long-term goals and objectives. We must evolve from sequential activities and "we/they" finger pointing to team effort and consensus management wherein "we" is all of us, and "they" are all of the problems that "we" need to solve.

Each and every individual must be committed to doing the right things right the first time--to strive for 100% conformance to the requirements. And we must recognize the quality as the complete corporate driver--as the symptom of attention to detail in design, purchasing, R & D, engineering, finance, marketing, selling, managing, and yes, even manufacturing--as the sum total of performance leading to the product or service being provided.

All of the above is most effectively accomplished in organizations with an open and participative management system where the emphasis is on positive reinforcement. Control for quality is then achieved through goal setting supported by an appropriate information and measurement system.

Our interest in quality should be motivated by a desire to reduce cost and, thereby, to increase profit. If we believe, with Lord Kelvin, that one doesn't understand a thing unless it can be measured, then to MANAGE any specific system element we must develop and implement a real-time, authentic information system to summarize and quantify the volumes of information and "feeling" into meaningful, concise action documents which assist in definition, prioritization, and follow-up.

Management (all levels and functions) needs to know in real time whether the system and its several elements are working and if not, why not? What corrective action should be taken? And after it was taken, did it have the intended effect? In the context of total or company-wide quality improvement, it is imperative that management reporting systems include the COSTS OF ACTUAL AND POTENTIAL DEVIATIONS FROM REQUIREMENTS as one of its critical elements.

Let us accept the premises: (1) that an open management style emphasizing trust, confidence and open communication is the best organizational structure to reinforce productivity, quality, and, in fact, all other positive business performance; (2)

that "quality" is the symptom of the sum total of performance leading to the product or service being provided; (3) that "quality" is simply defined as providing the customer what he wants; where he wants it; when he wants it, at the agreed upon price; (4) that measuring the symptom--the vital signs-the blood pressure and temperature--of all the performances that add up to customer satisfaction translates into somehow measuring quality performance. Then the question facing us is how to provide a reporting system to provide the management information to monitor performance; to chart progress and, thereby, to be able to provide the mid-course corrections to assure that all of the inputs are kept within focus to achieve continuing and ongoing quality improvement within the open management style discussed above.

In this context, I have found three reports to be extremely useful: the cost of Quality Report (COQ), the Numerical Quality Index of Customer Performance or the Customer Complaint and Rejection Report (C2R2), and the Product Review and Evaluation Program Report (PREP). The premise in using these three reports is first of all that PREP (Appendix A) provides a check-list--a discipline for the entire new product design and development, manufacturing and production cycle--that it gives all of the parts of the organization involved in the process a means of assuring that the required elements for a smooth new product introduction and for smooth technology transfer from design through manufacturing and processing, to customer delivery and after-delivery support, are coordinated and time phased so as to be most effectively carried out. This check-list can, of course, be amplified or extended into critical path timing charts. We have found the check-list useful in and of itself essentially as the summary sheet for all discussions, coordination meetings, etc., in support of the communication process. PREP is a means to the end. The other two documents: the Cost of Quality Report (COQ) and the Customer Performance or the Customer Complaint and Rejection Report (C2R2) are really the financial performance sheets for the internals (COQ) and the externals (C2R2). The Cost of Quality Report (Appendix B) facilitates the evaluation of quality in financial terms--putting quality performance on a profit/loss basis easily understood by any executive.

Just as the COQ is an internal performance measure, the C2R2 is an external performance measure. It is an attempt to capture and quantify the issue of customer satisfaction. This is easily accomplished by adding the number of customer complaints (i.e., the number of instances per unit time in which the customer advised the supplier of the existence of one or more defects but used the material in spite of the identified deficiencies) to the weighing factor w (e.g., w = 2 as assumed in an earlier paper, or 3 or 4) times the number of rejections (i.e., the number of instances per unit time in which the customer was unable to use the material and therefore returned it to the supplier or scrapped it by direction from the supplier).

The Taguchi Loss Function (which quantifies the cost of product or process variation) provides this measure in product and process design. The Variance Report (which quantifies and catalogues the difference between actual and standard costs) and Eddy-Current Analysis (which shows the cost of rework, repair and retesting which result from a lack of first run capability) serve production and operations management. And the Cost of Quality Report (which summarizes and analyzes the total cost of nonconformance) serves all management levels and function by (1) quantifying the extent of the cost due to non-conformance or "bad quality," (b) prioritizing (in a Pareto sense) corrective actions, (c) justifying expenditures in terms of return on investment (i.e., reduction of failure and/or testing costs), and (d) providing recognition for results/improvements.

The cost of Quality Report is a management tool that enables us to measure, in a direct profit and loss sense, the effectiveness of our efforts to improve management systems, design, manufacturing, etc., thereby basing "quality improvement" efforts on sound business judgments rather than goodness, motherhood, or morality.

The elements of the system being proposed (summarized in Table I) have all been used successfully by many leading companies. The methods are tried, proven, inexpensive to implement and profitable to maintain.

The implementation steps (Table II) require, above all else, a strategic, long-term focus and commitment on the part of senior management. This commitment must then be communicated to and shared by all individuals associated with the organization. For successful implementation of Company or Division Wide Quality Improvement (C/DWQI) it is above all necessary for senior management to recognize the need for C/DWQI and to take leadership in its initiation and implementation.

Table I

QUALITY SYSTEM ELEMENTS

1. MANAGEMENT & MOTIVATION
   CREATING THE CULTURE

   A. Policy and Execution

2. MANAGEMENT QUALITY REPORTING SYSTEM

   A. Real-time system

   B. Action documents for definition, prioritization and follow-up

3. MEASURE FOR QUALITY

   A. Cost of Quality
      1. Prevention, appraisal, internal and external failure

   B. Loss function for measure of product and process engineering design and effectiveness of purchasing activities and supplier base.

   C. Attention, prioritization, justification, recognition, and corrective action.

4. SHOP FLOOR QUALITY REPORTING SYSTEM

   A. Statistical Process control and Pareto Analysis in control chart form including implementation plan

   B. Closed loop and process improvements

5. PROBLEM SOLUTION INITIATIVES

   A. Problem Avoidance
      1. Product and process design of experiments including Taguchi methods, Contingency table, Multiple regression, Statistical tolerancing, simulation, Failure Mode & Effect, Analyses, etc.

   B. Problem Solving Management

Table I (continued)

6. <u>METROLOGY</u>

   A. Consistency in understanding the requirements
      1. Evaluation
      2. Execution

   B. Implementation

Table II

SIX STEPS TO TOTAL (COMPANY-WIDE) QUALITY IMPROVEMENT

```
____+____+____+____+____+_____ Time
```

PRODUCT   I.   DO IT, INSPECT IT, FIX IT

Gate-keeping. While we make improvements, let's try to minimize the amount of defective product that gets out the door.

PEOPLE   II.   CREATE THE CULTURE

Commitment, policy statement (development and implementation), strategy, communication, measure (including Cost of Quality).

PEOPLE   III.   IMPLEMENT CONTROL FOR QUALITY NOT QUALITY CONTROL

Provide the tools (including Education and Training)

PROCESS   IV.   CONTROL CURRENT PROCESSES

"Do the best we can with what we have" S.P.C. and Pareto in Control Chart form Emphasize CONTROL and operator responsibility.

SYSTEM   V.   PRODUCT & PROCESS IMPROVEMENT

Product and Process integration. Design for optimum product life. Continuous process improvement. Robustness against noise for LOWER COST.

CUSTOMER   VI.   QUALITY FUNCTION DEPLOYMENT

is the definition of the voice of external customer in operational forms throughout ALL FUNCTIONAL AREAS of the Company.

N.B.
All activities apply to external as well as internal suppliers, subcontractors, dealers, and after-service companies.

The main activities, after senior management commitment to and introduction of Company (or Division) Wide Quality Improvement (C/DWQI) methods will include, but not necessarily be limited to:

1. POLICY CONTROL

    Clarifying the mechanism of "policy control;" substantiating the upper stream of policy-making; securing consistency of policy deployment and strengthening "check" and "action" for the steady attainment of our company or division policy.

2. POLICY DEVELOPMENT

    Clarifying policy on quality and enriching the functional connections among purchasing, engineering, production, staff support, etc., in an effort to develop and produce a new product in anticipation of market requirements; and

3. WIDEBASED INVOLVEMENT

    Animating employee involvement and small group problem solving activities and expanding the same to affiliated companies and division (i.e., suppliers and customers) with a view to bringing out talents of employees and heightening their ability and interest in solving problems in a structural approach using Q.C. methods.

The important thing here is that the management philosophy necessary to foster C/DWQI is concerned primarily with the quality of people and their jobs. Virtually all Japanese companies and some U.S. companies have come to realize that quality products are more a function of employee knowledge and the way people interact than the application of traditional quality control technology. Prior to discussing the issue of MEASURE for quality it is important to recall several key points with respect to Quality Problems and Control for Quality vs. Quality Control.

CREATING THE QUALITY CULTURE

I think it is safe to say that every top manager, in fact, everyone involved in industry, wants to manufacture and sell quality products. The problem is that few understand how. In response to the questions "Who wants to do a bad job?" or "Who does not want to do a good job?", everyone expresses a personal desire to do a good job while questioning whether others are equally committed. The usual response is "I want to do a good job, but 'they' don't" or "'they' get in the way." I have asked these questions of thousands of individuals including clerks, technicians, unionized and non-unionized hourly workers, front-line supervisors, "middle" management, senior management, line, staff...and I've yet to met the person who WANTS to do a bad job. I've yet to meet "they".

The first thing we should accept is the fact that the vast majority of people do want to do a good job. However, execution requires that we: (1) know how a good job is defined--what are the requirements?; (2) know, if we deviate from these requirements or expectations, in what direction and by how much we deviate; and (3) have the tools, knowledge, education, training, etc. to get from where we are to where we need to be. If any of these is lacking, all the desire in the world will be insufficient to assure that we do a good job. And the corollary to this is that punishment cannot motivate us to do a job if we don't know what we are supposed to do or how to do it, or if we aren't capable.

It is important to note that defining requirements, providing tools, training, etc. are largely MANAGEMENT responsibilities. Then it is necessary to create the environment, (the Quality Culture) wherein people are encouraged to use (or at least not discouraged from using) the tools, training, etc.--another management responsibility.

As I discussed in an earlier publication (Transactions of the 36th Annual Quality Congress of the American Society for Quality Control, Detroit, Michigan, May, 1982, p. 21), quality products can then be achieved through reward--through positive reinforcement. Poor quality cannot be eliminated via punishment. This, I believe, is the reflection of behavioral psychology's fundamental principle: that positive behaviors are created and subsequently reinforced through reward or positive reinforcement while negative behaviors are extinguished through punishment. To change behavior via positive reinforcement is a slower process than to change behavior by negative reinforcement. To eliminate negative behavior is faster than creating and then sustaining new positive behavior. This long time frame poses great difficulties. It is very difficult, even when we accept that a long pull (perhaps ten years or more) is required to solve the problems, to feel that sense of urgency, that need to begin immediately, and to sense the urgency day-in-day-out, over a ten year period, when we will only make minuscule progress on any given day.

First and foremost, to operate successfully over a long time horizon, Deming points out (Automotive Industries, December 1981, p. 73), "top management people must understand their jobs...if they don't why just forget it." If top management does not understand their job, then the leadership and focus required will not be forthcoming and, in light of the long-term commitment which will be necessary, focus within the organization becomes paramount. In any organization the leader must provide the focus--must be committed to producing quality products. This commitment to quality as precursor to productivity results in certain organizational imperatives--specifically open communication, a people orientation and a zero defect philosophy to name just three. Moreover, this commitment to quality (as opposed to mere involvement) has specific consequences for all major elements of the leadership team (design, materials, manufacturing, and quality control) and their subordinates: respective responsibilities. A further result of this approach is that the Corporate Director of Quality Control is a person without a future.

Our previous approach of short-term goals and short-term financial incentives does not provide the kind of focus which is required for the new time horizons with which we must become concerned. Furthermore, this questions of focus is not merely one of asking that the chief executive officer know where it is we must go, but rather that every single person in the organization know the answer to that question.

William Ouchi (Theory Z, Addison-Wesley Publishing Co., Reading, Mass., 1981, Chapter 2, p. 39, et seq.) compares the American obsession for SPECIFIC, SHORT-TERM objectives with the Japanese desire that all employees understand the CORPORATE PHILOSOPHIES.

"The basic mechanism of control in a Japanese company is embodied in a philosophy of management. This philosophy, an implicit theory of the firm, describes the objectives and the procedures to move towards them. These objectives represent the values of the owners, employees, customers, and government regulators. The movement toward objectives is defined by a set of beliefs about what kinds of solutions tend to work well in industry or in the firm; such beliefs concern, for example, who should make decisions about what kinds of new products the company should or should not consider.

"Those who grasp the essence of this philosophy of values and beliefs (or ends and means) can deduce from the general statement an almost limitless number of specific rules or targets to suite changing conditions. Moreover, these specific rules or targets will be consistent between individuals. Two individuals who both understand the underlying theory will derive the same specific rule to deal with a particular situation. Thus, the theory provides both control over the ways people respond to problems and coordination between them, so solutions will mesh with one another. This theory, implicit rather than explicit, cannot be set down completely in so many sentences. Rather, the theory is communicated through a common culture shared by key managers and, to some extent, all employees.

"The organizational culture consists of a set of symbols, ceremonies, and myths that communicate they underlying values and beliefs of that organization to its employees. These rituals put flesh on what would otherwise be sparse and abstract ideas, and bring them to life in a way that has meaning and impact for a new employee. For example, telling employees that the company is committed to coordinated and unselfish cooperation sounds fine, but also produces skepticism about the commitment of others and creates ambiguity over just how a principle might apply in specific situations. When, on the other hand, the value of cooperation is expressed through the ritual of ringi, a collective decision making in which a document passes from manager to manager for their official seal of approval, then the neophyte experiences the philosophy of cooperation in a very concrete way. Slowly, individual preferences give way to collective consensus. This tangible evidence shows true commitment to what might otherwise be an abstract and ignored value."

The key result of this understanding and commitment by ALL is that attitudes, understanding of priorities and actual execution seem to be far more uniform in Japan than in the United States. In addition to attention to design, tooling, operating patterns, etc., there is obvious dedication, intelligence and cooperativeness among management, white collar and blue collar individuals all committed to working toward production of high quality parts. Also, attitudes, understanding of priorities and actual execution at the plant level seem to be far more uniform in Japan than in the United States. Efforts seem to be more methodical. Improvements appear to be more carefully planned.

The Japanese efforts toward quality improvements have "top-down" and "bottom-up elements as well as interdepartment, intergroup--"side-to-side"--elements. This goes beyond Total Quality Control (beyond integrating quality technologies into all functional departments--e.g., engineering, design, purchasing, R & D, finance, marketing, selling, managing, and yes, even manufacturing) to also include what Larry Sullivan (Company Wide Quality Control for Automotive Suppliers, Ford Motor Company, 1985) calls "quality function deployment" wherein the "voice of the customer" infiltrates the entire organization. This is perhaps most easily visualized by broadening our definitions of "suppliers" and "customers". We should think of the customer as anyone to whom we provide a product or service (whether internal or external to the company), and suppliers as all those who provide us with our inputs. Then we must meet the requirements independently of whether the customer is an internal or external one.

In the "top-down" approach to implementing their corporate strategies, the Japanese attitude of cooperation toward a joint goal rather than the U.S. tendency toward organization provincialism has led to an organization and an operation that is tailored to central analysis, priority and target setting and a great deal of dedicated, coordinated efforts toward achieving these goals. In contrast, in the United States (e.g., in the automotive industry) we have tended to decentralize operating control for quality and reliability to line organizations and have tried to give them the tools to control most elements on the basis that we can then hold them responsible for any quality inadequacies. The Japanese system is based on the concept of cooperation and improvement with no thought of assigning blame for failure. In the Japanese concept there is no such thing as individual blame. If something fails, the whole organization has failed. In contrast, our system is based on an assumed minimum need for cooperation in that each organization has explicity assigned objectives. I believe that the above is one of the keys to understanding why the Japanese system works so well and why the American system works so poorly by comparison.

The most important single contributing factor to Japanese successes in world markets must reside in their commitment to education and training all employees on a continuing basis. The most important assets of any organization go home at shift end. Continuing education and training coupled with long-term focus and commitment enable us to maximize the most powerful resource of any company--the human contribution.

With the recognition of the need for long-term focus combined with the requirement that all individuals involved understand their jobs, it follows that short-term measures and short-term goals may be inappropriate or counterproductive to the achievement of long-term goals, and may be in conflict with the strategic focus. It is necessary that we recognize that profits need to be optimized, rather than maximized. In this context, I would point out the very real difference between optimization and maximization by reminding you that an organization is very much like a human being. A human being whose temperature is optimized would find his temperature at 98.6°, whereas a human being whose temperature is maximized would find himself dead. W. Edwards Deming (Ward's Auto World, November 1981) states "that the obligations of top management can be boiled down to fourteen points; that these fourteen points not only cover what executives must do, but also provide a yardstick for anyone who wants to measure management performance." The first of his fourteen points is "creative constancy of purpose. The next quarterly dividend is not as important as the company's existence five, ten or twenty years from now." We must recognize that the historic lack of long-term commitment and strategic focus often derives from the combination of the relatively short tenure of management in management and the resulting short view generally taken by people who are used to outrunning their errors, and stockholders who demand short-term earnings per share performance.

Management has, in the past, paid undue attention to quarterly profits. This is typified by the observation that "long-term planning in the automotive industry is the executive decision what should I eat for lunch yesterday." If I focus on short-term profits, it is not certain that I will produce long-term profits. If, however, I focus on producing profits long-term, then after an initial period of investment which may be required I will be able to assure short-term profits. Short-term profits can, for example, be produced by minimizing maintenance, by shipping parts that perhaps should be scrapped, etc. The symptom of this policy is bad quality. The cause is a myopic view of profit--a maximizing of profit rather than an optimizing of profit. This approach is especially counterproductive if, as asserted in Business Week (March 12, 1979, special report, p. 32B), "Increasingly, quality is becoming more important to the customer than price, and making a product right the first time is found repeatedly to be less expensive than fixing it later."

When we speak of focus as something that must be provided by top management, we must recognize that this focus must be shared by all members of the organization, must be understood by all members of the organization, and all members of the organization must be committed to this focus. While it may be faster to dictate the commitment in the early stages, it is surely more advantageous in the long run for the commitments to evolve in a shared way. The Japanese point out that in the United States we do not reach consensus, but rather that a decision is made, an edict is issued, and then a wish exists that it will be carried out. One of the advantages of the consensus evolution of a shared focus is that there is then ownership of these objectives at all levels, leading to very real strategic goals rather than simple wishes.

More often than not, Orwellian double-think and double-talk further confuse the long-term efforts by the use of imprecise language. As proposed by John Groocock, Vice President-Quality, TRW, Inc. (private communication, 1983), we must carefully differentiate between the internal and external definition of "quality". From the customer's point of view, "quality" is getting what you want, when you want it, where you want it, at the agreed-upon price. A deficiency in ANY of these requirements is simply BAD QUALITY. The cause (excuse) is irrelevant. INTERNALLY the understanding of cause-effect is critically important, and hence the use of precise language is imperative. As Philip Crosby points out (Quality Magazine, August 1980, p. Q15)--referring IMPLICITLY to the INTERNAL definition of quality--there is no such thing as a quality problem. We must call problems by their generic names: design problems, plating problems, legal problems, etc. A quality problem means that the quality department has a problem. If that is not the case, we should say "non-conformance problems" or--even better--deviation from expectation. If we define a problem as a deviation from requirements and quality as conformance to requirements, then the symptom we call "quality" is merely the absence of problems in all elements of the organization since all elements contribute to the end product or service. Furthermore, any variation is potentially a problem to someone somewhere. One of our goals must be the understanding, then the control, then the reduction, then the minimization and ultimately the elimination of variation. If we do not recognize the requirements that we solve these problems in order to get at the symptoms, then we do not have adequate focus. We must first recognize (Deming, Iron Age, October d14, 1981, p. 61) that only "15% of all quality problems are related to a particular worker or tool. The other 85% arise from faults in the company's system and will continue until that system is changed."

Since we all want to do a good job, why aren't we more successful at designing in quality or engineering it in or even manufacturing it in? Because, first and foremost, we have to manage it in! That's the 85% of the problem. To overcome this, the main thrust of any quality improvement program must be that attention to quality comes first: first in the preproduction planning cycle, first in the design cycle, and first in the manufacturing cycle. Since the key to eliminating "quality problems" is to attack them at their source (rather than try to correct them after the fact), it is important that we look at the pivotal role of the design engineer and the process of designing a part or component (or process). Again, we must discuss not only the extinguishing of bad previous behaviors in design, but also the creating of new behaviors. For example, in the automotive industry people were used to dealing with fairly large pieces. If a problem arose, they simply added more material--upsized if you like--in the hope that this upsizing would then permit us to gain added strength, rigidity, or whatever, and as a result reduce the so-called quality problem. This luxury of upsizing, or upstrengthening parts is no longer available to the design engineer as a knee-jerk reaction to "solving quality problems". In addition, the discipline which we refer to as "new product design and development" must be changed.

Historically, the design of a particular part has proceeded roughly as follows: the design engineer designs the part and then takes this predesigned part to the materials specialist to look at the proposed material to find the closest "real life" material to do the job within economic constraints. Invariably, the material postulated does not exist at a reasonable price so a compromise is made. The part now is turned over to manufacturing in order to process the part according to the best available techniques, and finally, after the fact, some control procedures are added. In this approach each part of the system is optimized, given all that has gone before, rather than optimizing the entire system with appropriate sub-optimization of

17

each of the sub-systems. For example, it may be possible with a minor change in design (perhaps something as simple as a change of radius) to greatly simplify the manufacturing and control process and, thereby, make a part which is functionally as good as or better than the original part, but which is much easier and less expensive to manufacture and control. Furthermore, since it is the design engineer who must accurately knows where the design corners were cut, it should be the design engineer who has major input into defining the control system. To have the manufacturing people along determine what the quality control system should be is much like the drunk who loses his wristwatch in a dark alley, but looks for it under the street lamp because it is so much easier to see there. What I am proposing is that we look at the entire design, materials, processing and control system as a four-legged table and adopt a systems approach where one sub-optimizes each of the individual systems in order to optimize the entire system. A simple change in design might result in enormous simplifications in manufacturing. Knowing what the critical design parameters and/or the critical manufacturing process parameters are would permit much more cost-effective reliability assurance on the part. This systems or team approach, rather than a sequential process of decision making, results in a greater commitment on the part of all concerned to those decisions in which all have contributed and had a part. It requires that the design engineer be "first among equals" in order to provide the focus, but it also means that all of the members of the team trust and count upon the other team members to do their jobs. This makes the design engineer much more a part of the leadership team, perhaps even the lead element in a consensus approach to design while, at the same time, assuring that the remaining elements (materials, processing and control) are committed to the design once the ENTIRE team has completed its work. This results in a commitment to the whole, rather than a commitment to a particular piece of the whole and a finger-pointing involvement in the rest.

What organizational structure will maximize positive feedback, management focus and the encouragement of long-term commitment in order to contribute to producing an optimum quality product? With this question in mind, let's examine issues of organization of quality assurance or reliability assurance functions in particular with emphasis on the question, "Organizationally, what is the best approach to assure a high quality product for the customer?" It is useful to start this discussion by looking at the Japanese and asking what does "Made in Japan" mean to you? Around World War II, "Made in Japan" was synonymous with junk. Today, "Made in Japan" is synonymous with quality, with high technology, with high reliability, high quality products and , most of all, high customer satisfaction. Instead of junk, when someone today says "Made in Japan" we think of Sony, Toyota, Datsun, Mitsubishi, etc.-- companies with a high product-quality image. The once derisive term "Made in Japan" today signifies unparalleled standards of quality and reliability.

An excellent article by Peter F. Drucker entitled "What We Can Learn from Japanese Management" (Harvard Business Review, March-April 1971, p. 110), references as major contributors: decision by consensus, lifetime employment, continuous training, and the godfather system. Other authors stress equally significant contributors to the so-called Japanese phenomenon. We must recognize that assembly-line technology exists in the United States; it exists in Japan. Assembly lines are generally not conducive to a worker's social integration in the factory. Yet, somehow, in spite of this apparent absence of job satisfaction on an assembly line, the Japanese found a way to obtain a high commitment from their workers. They have found a way of imbuing their workers with a sense of belonging, a sense of identification with the company, a sense of social integration, and as a result of this have been able to achieve a high degree of productivity, quality and reliability. They have done this by several key actions. First, there appears to be total inclusion of the employee into the work organization. The approach is collective, not individual. it includes the entire family, not just the individual and applies to personal and work-life and especially to the responsibility for quality. In addition, there seems to be an extremely high identification of both the individual and his family with his company and peers. The employee feels a sense of fairness and a great deal of security provided to the employee by the employer. There also seems to be a great degree of openness; a great amount of feedback in many directions--feedback from subordinate to superior, feedback among the various functions (design, materials, production and control), feedback between the supplier and customer, etc. In other words, mutual trust and openness seem to be much greater in Japan than in the United States. For example, the exchange of warranty and policy data as well as all other field information between customer and supplier in Japan is much greater--in fact, surprisingly greater--than in the United States.

Another thing that characterizes the Japanese approach to obtaining and maintaining high quality is that high quality is rewarded. Poor quality is not penalized. The assumption is made that no one would deliberately produce poor quality. If someone produces poor quality, or finds poor quality that a different individual produced, he will fix it. He will fix it because the assumption is that quality is everybody's business. The assumption is that no one is producing poor quality deliberately and, hence, you do not point fingers at anybody by taking appropriate action to improve quality. You are merely assisting. This is why line stop buttons exist along manufacturing lines in Japan. Any operator can stop the line to correct a defect. This approach is still rare in American industry. Another item that categorizes the Japanese industrial climate is that a great deal of attention is paid to continuous training (producing generalists rather than specialists) with age and seniority being much more important in Japan than in the United States.

An interesting Japanese approach called Quality Control Circles has received much attention in the United States and has become almost single-handedly credited with the bulk of the rationalization of the Japanese success. I believe that Quality Circles are a critical derivative element of the Japanese system which flows from the assumption that no one would deliberately produce poor quality and from the management attitude and focus provided in Japanese industry. For anyone wishing specific details concerning Quality Circles, I would cite "Quality Circles--An Annotated Bibliography" (Quality Progress, April 1981, p. 30) which provides an excellent extended bibliography on QC Circles. Suffice it to say for our purposes, QC Circles are voluntary organizations made up of personnel from all levels of the company, though mostly made up of production workers. These QC Circles' main purpose is to facilitate training and developing workers. In a QC Circle, the supervisor will generally teach his subordinates. How many supervisors in the United States, at any level, are willing or able, in a formal sense, to teach specific problem-solving techniques to subordinates? Since the assumption is made that no one would deliberately produce poor quality, it is important (and the QC Circle is one major mechanism for accomplishing this) that each employee have the tools to be able to recognize what is currently being produced, what needs to be produced, and how to get from point A to point B.

The Japanese first ask whether defects found are operator controllable. Does the operator know what he should be doing, and can the operator do what he should be doing? That is to say, is it possible to regulate the process? Those are the questions that must be asked. These are the tools which must be provided to the operator by the quality control director (rather than merely serving as an inspection police force). What should the operator be doing? What is he doing? And, is he capable of controlling the process? The operator must be able to answer these questions if he is to look at the quality of his work and control the parts that he is producing, taking corrective action as required to assure high quality production.

If a participative open style of management exists within a corporation, the QC Circles appear to be a natural extension and a good way of minimizing we/they conflicts involving employees and reinforcing confidence and participation. Without an open style of management, the QC Circle is doomed to fail and be perceived as mere window dressing. However, by opening regular lines of communication, encouraging workers to express themselves and considering workers as individuals, a Circle-type program can give a tremendous psychological lift to those whose self-esteem begins deflating as they drive into the parking lot and, most importantly provide the substantive tools necessary to those production workers. The critical point here is that a QC Circle will be beneficial if and only if line management has been give, AND HAS ACCEPTED, "ownership" of the Quality Control Circle. This ownership is manifested by a management style, a knowledge of job and a willingness to teach that job to subordinates. In a very cogent article in the Automotive Division Newsletter of the American Society for Quality Control (May 1981), Hans J. Bajaria points out that while it is clear that the United States has a problem, it is not clear whether QC Circles will provide appropriate and significant solutions to that problem. Further, he states that before valuable resources are invested in quality control programs, certain key questions concerning the dominant types of problems to be solved and management attitudes already in existence need to be answered.

Another key to providing shaping-feedback is that the inspection function is assigned to the production worker. Now, think about that. The inspection function is assigned to the worker. In the United States, we would say that is akin to having the cat watch the cream, and yet the proof of the pudding is that it works. What is critical here is that the initial assumption, that given the opportunity to produce quality products the worker will do so, yields a set of actions which enable the worker to live up to this expectation. On the other hand, the assumption that the worker will either deliberately or absent-mindedly screw up whenever allowed to--from that assumption flows the need for a police force (an independent quality control inspection force). Making the production workers responsible for quality enables him to build it in, whereas the quality control inspector can only test and this testing does not improve quality; it simply reports to the worker something he could have determined for himself. Why not eliminate the extra individual--the policeman with whom the worker will develop an adversary relationship? If the production worker does his own testing, he reports the need for corrective action to himself.

Though I cannot quantify it, I am quite confident the quality of work produced by a person who is responsible for his own actions is greater than by those who simply say: "Why bother? My supervisor or QC inspector will discover the errors. I don't need to be as careful." It is much like the situation that occurs when I drive my car alone vs. when I drive with my wife in the car. Now my wife has got to be (and I must be fair to her) the best backseat driver ever. When she is in the car, I will depend on her. I don't do it consciously. I don't do it deliberately. I don't deliberately drive carelessly, but I know that she is going to watch the road. She is going to watch out for other cars. She is going to warn me. I, therefore, at least subconsciously relax. When I drive alone, on the other hand, there is no one to help me. There is no one to warn me. Therefore, I find that I am more diligent, more alert.

When the Japanese speak of "them and us" they speak of "them" as all of the problems that "us" have to solve. In America when we speak of "them and us", more often than not, "us" is the particular group with whom we are involved (whether that be design or processing or materials or control--it depends on the function in which we find ourselves), and "them" are all the other people involved in the other groups. Once it is accepted that "them" are the problems that need to be solved and "us" are all of the people that need to solve them, and the other assumptions concerning people's attitudes are made, it is then necessary to train people to identify and analyze, solve and correct the problems. The Japanese, therefore, help operators solve their problems, teach operators how to hold the gain once these gains are made and accomplish this by the use of modern statistical control methods.

The combination of open style and a focus on more strategic matters is, in fact, a devastating one for competition. This approach is based on the assumption that people in general are anxious to do a good job; that they are self-motivated and that given enough information)focus) to know what is expected, to know what a good job is, how it is defined and how to do it, that people in general will make every effort to do a good job. Theory Y management philosophy requires good communication channels, up, down and sideways throughout an organization. It not only requires but encourages participation on the part of all employees. Appropriate awards, incentives and recognitions should emphasize group, rather than individual, performance.

Hancock and Plonka reported in a SAE paper in 1977 on the following experiment: In an automotive assembly plant, special teams were set up in order to improve quality and sure enough, these teams did an outstanding job of improving quality. The teams were then removed--eliminated--and quality immediately deteriorated to the pre-experimental level. The next experiment was to involve line personnel in systems development, rather than set up separate special teams to implement and monitor improvements in quality. Lo, and behold, with the involvement of line personnel in systems development, quality went up, production costs went down and the gains were held over long periods of time--periods of time which exceeded the "duration of the experiment", or more specifically, exceeded the duration of specific management interest in the experiment. This experiment, I believe, highlights the difference between authoritative and participative management and shows quite dramatically how participative management creates reliability awareness; how by an open management style we maximize communication, maximize subordinate involvement in the goal-setting process, motivating through involvement rather than fear, and sharing the credit with those who are most capable of and most necessary in solving problems. Product reliability, therefore, seems to be intimately involved in the choice of a participative form of management. The greater the degree of participative management, the greater the probability of a high product reliability. Production people, design, manufacturing, services--all phases--are committed to, not merely involved in, the product reliability decisions and actions and, therefore, are more committed to producing a high quality product.

In a participative management organization the role of the quality control manager or director is much changed from what it is today. Juran points out that the role of the QC Director must be greatly expanded. The role becomes not one of the policeman, but rather of educator. It is the function of taking the equivalent of QC Circles an teaching people to analyze problems, to solve problems, teaching an awareness, obtaining a knowledge of the needs people have and, thereby, assuring a high quality product. It means that the corporate director of quality control and reliability engineering should report to highest management levels and be given, if not positive leadership from management should report to highest management levels and be given, if not positive leadership from management then sure, strategic focus. With attention to detail required at all levels and a more open, responsive, communicative, and participative form of management, the design engineer must anticipate misuse in order to reduce product liability exposure and must, in fact, accept more responsibility. Marginal design alternatives for lower initial cost invariably add to product life cycle cost. Somehow this must be monitored to assure that these kinds of short-term alternatives are not chosen. The only people not truly responsible for quality in an organization are the quality control personnel since they are, in fact, the only department with no real control over the product either in design, in manufacture, in materials, or in procurement. Since reliability and productivity are everyone's business in the corporation, rather than have a separate reliability department, reliability responsibility must rest with engineering, design, manufacturing, accounting, sales—in fact, every part of the organization. What this means is that the quality control professional, the director of quality control and reliability engineering, must function as an organizer, an administrator, a policy maker, an expedititor, an educator, but not a policeman. What this also means is that more tools must be provided for the designer, the manufacturer, etc., in order to assure product quality.

In the future, I suspect that the quality control department as it now exists in Most American corporation will disappear. It will be replaced by a reliability function (probably a staff function) advisory to the chief operating officer on policy, and will set policy but not be a policeman. The new manager or director will find himself or herself concerned with such items as finance, marketing, manufacturing, production, design, product liability, legal issues and all of the other strategic corporate issues which are, after all, the causes of or contributors to the symptom called quality. These kinds of demands on the reliability assurance director are going to require a much greater background, a much greater knowledge of all phases of the company's activities on the part of that director. The quality control director, as we know him today, has an extremely limited future. Quality, like planning, must be integrated into all operations to assure commitment to execution.

J. M. Juran (Management Review, July 1981, p. 57 points out that in the West, it is common for senior manager to be trained in how to use budgets, balance sheets, profit statements and financial controls of all sorts. Most senior managers use these devices. They participate in formulating financial goals, in reviewing results against goals and in taking action based on these reviews. Through these means, senior manager maintain effective leadership over their company's financial performance. A critical question to be answered in the '80's is how to gain for senior management the same effective leadership of their company's quality performance.

SOME THOUGHTS ON CUSTOMER-SUPPLIER RELATIONS

Transition periods when industries, economies or societies go through rapid changes, are generally periods of unparalleled creativity. In order to harness this creativity and accomplish the requisite redirection of "smokestack America", it will be necessary for certain changes to occur in the way that business is conducted.

Previously, the various regional industries have operated in the comfort of: (a) static, stable, relatively undifferentiated domestic product protected by various barriers to competition from other regions and different products; (b) stable workforce; and (c) a small cadre of trained people to organize and control this hard-working, loyal and dedicated semi-skilled workforce by rote methods.

In this static, protected, constrained environment, both the customer and supplier adopted a short-term focus and commodity strategies typical of industries which view themselves as "mature" and producing a "mature product".

Following this ear of stable products, relatively stable workforce, short-term focus, commodity strategy, a mature product and a requirement that we "run lean", basic industry was forced to undergo an intensive metamorphosis characterized by sales erosion, tightening of quality standards, high inflation, tight money, deep recession, products which are no longer unique, a dynamic materials environment, a compressed time frame, and other "radical" upheavals.

In a society becoming visibly more consumer oriented, products and services are being tailored to (a) more closely satisfy the customer's specific and varied requirements, and (b) provide a wider range of real product and/or service alternatives. To illustrate the pervasive and general nature of these changes let us consider, as example, the banking industry.

Previously the banking industry offered a limited choice of roducts--"savings" banks offered passbook savings accounts while "commercial" banks offered checking accounts. Customers moved from one bank to another to take advantage of small differences in interest rate (on savings) or small differences in cost (on checking accounts). A static, commodity product in a mature industry begat a dynamic customer-supplier relationship just as it had in the auto industry!

Now, a specific bank attempts to offer a complete "umbrella" of products and services. The product mix--the range of real alternatives--is greatly expanded and includes such investment vehicles as certificates of deposit, time accounts of varying duration, interest-bearing checking accounts, etc. The supplier (the bank) attempts to provide a dynamic mix of products and services in the hope of encouraging a long-term (static) customer-supplier relationship.

My first thesis, simply stated, is that a static (stable) commodity product in a mature industry begets dynamic customer-supplier relationships whereas a dynamic, differentiated product in a demantured industry will tend to encourage more sable, longer-term supplier-customer relationships.

It is expected that anywhere from 5 to 10 years will be required before significant inroads will be made and consolidated in the area of quality improvement. This combination of dynamic environment and longer time horizon than we are used to in industry highlight the need for planning, for focus and for dedication. (Planning is not considered to be as critical in a stable environment as it is in the current more dynamic environment).

All of the major partners in industry (final assembler, supplier, labor, government and financial institutions) will need to develop new ways of operating in order to survive and prosper in this new, more dynamic environment. AS pointed out by Edward Veal suppliers must be judged on consistent quality and overall contribution to the job at hand; must demonstrate a commitment to quality/productivity methods; must demonstrate a capacity for technical innovation and help solve design and product performance problems for their major customers; must have the size and financial strength to make investments in new or improved processes and support research and development efforts through long lead times before new technologies result in business returns; must have strong business planning and a hardnosed attitude toward evaluating new opportunities; and must be more diversified and thereby balance automotive risks and rewards with more stable business segments. The result of this profile will be fewer, larger and more involved suppliers. There also will need to be more cooperative supplier-customer relationships at all levels.

The new era supply base will be a partnership in which customers (primary assemblers) will have the major coordinating role. In their own interest they will do several things:

- They will provide a basis for stability with suppliers by forming longer term relationships on specific new technologies in which suppliers make investments and take risks. The particular shape of these arrangements will vary, but certainly will include long term supply contracts.

- They will make purchasing decisions based on overall supplier appraisals that put quality first.

- They will allow prices that afford good suppliers an opportunity for long term profitability. (Suppliers who do not recover their investments in new technology and facilities soon will put their resources elsewhere.)

- They will refrain from integrating areas where a good innovative supply base exists. This is just good business sense. Capital requirements are too great to allow manufacturers the luxury of duplicating supplier facilities.

- They will not only invite but insist on early supplier involvement in new programs to insure the best and most cost-effective designs.

The forces are in place that will bring about the changes described. Certain suppliers are committed to advancing technologies in which they are involved, exploring new ones and providing new products. The degree to which we are all successful in bringing about an industry renaissance depends on our capacity for well thought out change.

William Agee points out that we (suppliers) are walking the same economic tightrope as the OEM's. Like the OEM's, suppliers will be required to make huge capital investments in the next 10 years. Since most suppliers are relatively small, some will not have the resources for the needed financial commitments. The cost of capital is the same (or higher) for the supplier as for the OEM and both must be cognizant of the demand placed on the other. The "shakeout" continues in the OEM's and similarly the number of suppliers is dwindling, so a mutual understanding of the precarious economic tightrope each must walk in these difficult times would be extremely beneficial to supplier and OEM alike. For instance:

1. Suppliers must become much more innovative on their own, helping their customers develop product at the lowest possible cost. And the OEM's might be well served in encouraging their suppliers to promote and develop new ideas.

2. Considering the climate we face, the OEM's should recognize the high risks as well as the significant investments required in the development of new products. The supplier, in order to justify these investments must be assured to long-term market commitments.

3. The traditional "market test" which asks suppliers for quotations should be revised. As presently constituted, the system rewards short-term commitments. It is wasteful and fails to promote innovation, productivity, and quality improvements.

Mr. Agee concluded, "overall, I recommend that OEM's sit down with suppliers and discuss their needs, taking advantage of facilities, other resources, and a mutually acceptable long-term strategy. Adhering to short-termitis will not benefit either party."

Relations between customers and suppliers need to change. Suppliers will need to supply parts and technology. Since there are up-front costs to be born by suppliers the customer cannot simply say to the supplier, "Take the risk" without somehow being prepared to provide (or at least allow) for adequate return on investment in order to make the risk/reward balance acceptable to the company taking this risk. This, in my view, will require a very different commitment on the part of the customer to the supplier and conversely. Suppliers who provide technology and quality products with decreasing overall system cost must be rewarded by the OEM's with purchase agreements that encourage/permit (not guarantee) long term profit.

To discuss what suppliers can do, or should do, or must do, without discussing quality and productivity and their interrelationships would be to overlook the most important single item that suppliers can do for themselves. However, this was discussed previously and will not be repeated here.

This historic lack of sensitivity and commitment between suppliers and their primary customers and the attendance lack of mutual understanding and realization of the equity that each has in the other must be replaced by a new era of "supplier involvement" (similar in many respects to employee involvement). The fundamental lack of trust between partners (suppliers and primary manufacturers) is quite similar to the historic lack of trust between labor and management, and therefore may be improved/changed by the same kind of positive reinforcement, new dialogue and mutual understanding which is building between management and labor. To continue treating the supplier as the hourly or marginal worker in the industry will of necessity be mutually destructive for final assembler or manufacturer and supplier alike. The new dynamic stability which the industry faces will demand that a working relationship be built based on trust and mutual understanding. This new relationship can only be built based on positive reinforcement and a more open and honest relationship.

The trend in industry in the future among suppliers will be buy-outs, take-overs, associations, mergers, and failures. This will leads us toward a "two-tier" supplier system. Along with the large suppliers-the Eatons, Borg-Warners, Danas, Textrons, Rockwells, TRWs, United Technologies, Ex-Cell-O's and FM's of the world--will be found other large companies which will be added to the list via the merger, buy-out, take-over, or association route. And there will be a second tier in this new supplier organization. The innovators (who provide technology and solutions to problems rather than merely hardware) will have to be companies large enough to afford the substantial up-front investment and long pay-back periods. The "second" tier will tend to be the 'me-too' companies (which very well may be secondary suppliers to those primary suppliers). These companies will generally be smaller and less financially independent and will supply the less complex, commodity kinds of products. In general they will operate closer to the margin and be mainly suppliers of hardware rather than problem solvers and/or technology developers. In fact the structure will be MULTI-TIER, but the tiers will tend to be discrete, identifiable, small in number, and differentiated on the basis of financial ability to develop technology.

In this evolution of supplier-OEM relationships the traditional "market test" will need to be altered since, as presently constituted, the system rewards short-time commitments. Suppliers who provide technology as well as quality products while decreasing overall system costs must be rewarded by the primary manufacturer with purchase agreements that encourage/permit long-term profit.

In the context of a dynamic environment, a capital intensive industry and an uncertain (and declining)labor market, local governments are being asked to accept responsibility for creating a consumer market and business climate which will provide stability and allow local manufacturing to regain (or gain) international competitive position. Since our industrial world thrives in a favorable political climate with rising real consumer income, it will be necessary for government to help provide the climate to support these conditions. Independent of what form that "assistance" takes, we must at least agree that an adversary relationship between the governed and the governing is counterproductive and intolerable. The same kind of positive reinforcement and development of trust and confidence that is recommended as the major requirement in improving customer-supplier relationships and in improving management-labor relationships applies equally well to improving government-governed relationships.

Policy, planning and strategic focus combined with mutuality of effort are required. In face, historically, there has been a dual--industrial/governmental--aspect to success in worldwide automotive competition.

As a final point in this section, I feel it is critical to question how industry will attract dynamic new employees who can provide the focus, commitment, innovation, and creativity that is so desperately required? We must be able to attract and keep far-sighted business leaders. Ed Cole of General Motors pointed out upon his retirement that the auto industry in not fun anymore. Unfortunately, it is the fun business that provides the challenges that attract the dynamic business leaders. We have for too many years seen too many gray flannel suits in the executive suites of industry-- too many who are short term finance oriented, too few who dare to be great and too few who are willing to champion the need for high technology in a "mature industry."

SYSTEM, MEASURE AND CONTROL

Let us accept the premises: (1) that an open management style emphasizing trust, confidence and open communication is the best organizational structure to reinforce productivity, quality, and, in fact, all other positive business performance; (2) that "quality" is the symptom of the sum total of performance leading to the product or service being provided; (3) that "quality" is simply defined as providing the customer what he wants, where he wants it, when he wants it, at the agreed-upon price (i.e., 100% conformance to all of the requirements; (4) that measuring the symptom-- the vital signs, the blood pressure and temperature--of all the performances that add up to customer satisfaction translates into somehow measuring quality performance. Then the question facing us to how to provide a reporting system to provide the management information to monitor performance; to chart progress and, thereby, to be able to provide the mid-course corrections to assure that all of the inputs are kept within focus to achieve continuing and ongoing quality improvement within the open management style discussed above.

In this context, I have found three reports to be extremely useful: the Economics of Quality Report (EOQ), the Numerical Quality Index of Customer Performance or the Customer Complaint and Rejection Report (C2R2), and the Product Review and Evaluation Program Report (PREP). The premise in using these three reports is first of all that PREP provides a check-list--a discipline for the entire new product design and development, manufacturing an production cycle. It gives all of the parts of the organization involved in the process a means of assuring that the required elements for a smooth new product introduction and for smooth technology transfer from design through manufacturing and processing, to customer delivery and after-delivery support, are coordinated and time phased so as to be most effectively carried out. This check-list can, of course, be amplified or extended into critical path timing charts. I have found the check-list useful in and of itself essentially as the summary sheet for all discussions, coordination meetings, etc., in support of the communication process. PREP is a means to the end.

The other two documents: the Economics of Quality Report (EOQ) and the Customer Complaint and Rejection Report (C2R2) are really the financial performance sheets for the internals (EOQ) and the externals (C2R2). The Economics of Quality Report facilitates the evaluation of quality in financial terms--putting quality performance on a profit/loss basis easily understood by any executive. It is an outstanding management report useful for awareness, prioritization and justification (as we will see shortly).

With the recognition of the need for long-term focus combined with the requirement that all individuals involved understand their jobs, it follows that short-term measures and short-term goals may be inappropriate or counterproductive to the achievement of longer-term goals, and may be in conflict with the strategic focus. It is necessary that we recognize that profits need to be optimized, rather than maximized. In this context, I would point out the very real difference between optimization and maximization by reminding you that an organization is very much like a human being. A human being whose temperature is optimized would find his temperature at $98.6°$, whereas a human being whose temperature is maximized would find himself dead.

Let's carry the parallel between an organization and a person just a bit further. We don't deliberately get sick then go to the doctor for cure. (We don't operate in a do it, test it, fix it mode.) Nor do we stay healthy merely by the doctor monitoring our vital signs--temperature, blood pressure, etc. (i.e., we don't manage our lives merely by reading the P & L statement!). The doctor helps us stay healthy by EDUCATING us with respect to diet, rest, exercise, etc. Then we are responsible for our own actions. No QC police function here! We do the right things then verify that all is well by periodically measuring the symptom (temperature = $98.6°F$). We would consider it ridiculous to try to stabilize our body temperature at $98.6°F$ by surrounding ourselves with a temperature controlled bath or box rather than treat the virus that caused the $104°F$ fever! We treat causes and measure symptoms--we engage in control for quality, not quality control. In a problem solving mode we determine whether there is a deviation from requirements ($104°F$ vs. $98.6°F$) then find the cause (virus) then the appropriate medicine (corrective action), then take the medicine (implement the corrective action), then verify that the pill did the job (temperature = $98.6°F$)! We do the detailed things that need to be done then verify that we did the right things. We measure some symptom or parameter. We don't manage quality. We manage FOR quality.

1. Why measure?

   There appear to be only four reasons for ever making a measurement.

   (a) to characterize (e.g., as in P & L or new design).
   (b) to define a problem (i.e., to ascertain the magnitude and direction of the deviation from the requirements defined in (a).
   (c) to verify efficiency of corrective action (did the corrective action in fact take us from (b) to (a)?).
   (d) for defensive reasons (we live in a litiqeous society).

   MEASURE BY ITSELF ACCOMPLISHES (CORRECTS) NOTHING! The culture (MANAGEMENT) determines success!

2. Is the process controllable?

   (a) Does the operator know what he or she is supposed to be doing? (Does the operator know the requirements?)
   (b) Does the operator know what he or she is doing? (Does the operator know how, by how much, and in what direction actual deviates from requirements?)
   (c) Can the operator regulate the process? (Can the operator get from "IS" to "SHOULD BE"?)

   Items (a) and (b) require both a desire and ability to make problems visible and solve them. Item (c) is an issue of equipment, education, training, etc. ALL ARE MANAGEMENT OR CULTURE ISSUES!

3. What should we measure?

   Since measure is expensive, we should emphasize (in the context of item 1 above) those areas which present major opportunities for improvement (i.e., significant deviations from requirements). In the context of product (or process) engineering, the so-called bathtub curve of failure (Figure 1) is particularly significant in separating measure for design vs. process information.

   It is useful to think of the bathtub curve as having three regions:

       I. Early or "premature" failure--usually a result of some manufacturing or assembly deficiency
     II. Steady state or random failure
    III. Product design life failure made up of two Weibull distributions.
         A. Process failure
         B. Product design failure

We then can focus on either of the distributions comparing "actual" to "should be" in order to obtain authentic, real-time information upon which to base action.

IF WE ARE NOT PREPARED TO MANAGE THE SYSTEM AND TAKE ACTION TO ELIMINATE CAUSES OF SPECIAL VARIATION THEN WE WASTE TIME, MONEY AND EFFORT IF WE MEASURE SOMETHING JUST FOR THE SAKE OF MEASURE. Measure for measure's sake is absurd!

The first thing we should accept is the fact that the vast majority of people do want to do a good job. However, execution requires that we: (1) know how a good job is defined-what are the requirements?; (2) know, if we deviate from these requirements or expectations, in what direction and by how much we deviate; and (3) have the tools, knowledge, education, training, etc. to get from where we are to where we need to be. If any of these is lacking, all the desire in the world will be insufficient to assure that we do a good job. And the corollary to this is that punishment cannot motivate us to do a job if we don't know what we are supposed to do or how to do it, or if we aren't capable.

It is important to note that defining requirements, providing tools, training, etc. are largely MANAGEMENT responsibilities. Then it is necessary to create the environment wherein people are encouraged to use (or at least not discouraged from using) the tools, training, etc.--another management responsibility.

PROBLEM PREVENTION/PROBLEM AVOIDANCE/PROBLEM SOLVING

The most effective way to reduce the cost of quality is to apply well known and proven preventive methods prior to (i.e., upstream of) manufacturing to identify potential deviations from expectation (or deviations from requirements). This activity should be part of all functions of the company but must be a major part of the design phase of product as well as process development. Once the potential causes for deviations from requirements are identified and prioritized, then appropriate action should be taken to:

Prevent Problems

    Prevent even the possibility of occurrence by eliminating the cause.

Avoid Problems

    Minimize the probability of occurrence and/or severity of the deviation if it does occur. In this instance we cannot eliminate the cause but we can design for robustness against noise. In this category we will not consider the possibility of merely not using the product (scraping it) or stopping the process to "avoid" a problem. (This would be killing the patient to avoid his dying from the disease.)

Solve Problems

Solve those problems which we did not or could not prevent or avoid. In this case a cause yielded a deviation from requirements and, after the fact, we must take action to reduce the deviation by reducing the cause and/or its effect.

While people seem genuinely interested in solving problems the "we/they" attitude which often pervades organizations frequently gets in the way as does the fear, at the operator level, or reporting the full extent of the problems. Problems must be made visible (defined) before they can be solved. Granted one has to recognize that the effort required to solve some of our more vexing problems may be great. Nonetheless, it is important to overcome the tendency to "symptom solve", and to grab for the quick fix that will get the problem off our back (and thereby assure that it will likely reoccur).

We need to have certain elements in place to assure both effective problem avoidance and problem solving. These elements include above all else (first and foremost) dedicated and effective leadership especially with respect to:

A. Creating and Sustaining the Culture

Creating an environment within there exists the mentality or mind set that makes everyone <u>want</u> to make problems visible, analyze them (cause-effect) and solve them.

B. Project Control

Managing and coordinating responsibility for the various efforts especially (but not limited to) those of multi-disciplinary or cross-functional problem avoidance or problem solving teams.

C. Sense of Urgency and Tenacity

Prioritizing, allocating resources and maintaining a sense of urgency to solve problems at (or as close as possible to) the point of origin. (This includes the discipline of not jumping from problem to problem--of not changing priorities/direction--in order to assure that efforts, once begun, are brought to final/permanent solution and that problems get solved and stay solved).

D. Resources

Providing the necessary resources.

E. Education and Training

Providing the requisite education and training including (not <u>only</u>) Statistically Process Control and Statistical Problem Solving (but, in addition) Design of Experiments, Failure Mode and Effects Analysis, and the other disciplines previously discussed.

F. SPC

Implementing Statistical Process control and corrective action so that we bring stability to current processes by removing some of the out of control points (remove/correct/eliminate/reduce special causes of variation).

(As pointed out by Hans Bajaria (Reference 3), it is critically important to understand that STABILITY does NOT mean ABILITY. We can bring a process from unstable to stable with problem solving activities. That is the relatively or comparatively easy part of the improvement process. To go from STABLE to ABLE usually requires a much more consuming problem solving effort. But these latter efforts should be expected to yield savings--return on effort--return on investment--commensurate with the effort. Activities in this area should be more than justified).

Quality is itself a measure of how well we are doing all of the things we need to do to provide the product or service. And assuming we are serious about succeeding/surviving, it is critically important that we have an objective, quantitative, authentic, real-time set of measure rather than discuss quality as a degree of "goodness" or "morality."

We need to put quality on a quantitative basis related to P & L similar to operations and financial performance reporting. Cost based measure is useful for decision making purposes since it is a "universal", well understood language.

Simply speaking, our interest in quality should be motivated by our desire to reduce cost and thereby, improve profit. This must be accomplished by managing all of the details of the business.

COST OF QUALITY--QUALITY PERFORMANCE MEASUREMENT SYSTEM

General Description--Purpose

Our interest in Quality is (or should be) motivated by our desire to reduce cost and increase profits. This requires that we place quality on a Profit and Loss basis and develop a real-time management indicator to track performance. While the issue of Quality Cost is an integral/pivitol part of the Quality Management Reporting System, it is deemed to be of sufficient importance in the overall improvement process to be treated separately.

It is important to note at the outset that our interest in Quality Costs must include both off-line activities (such as product and process engineering) and on-line operations (such as manufacturing). The more we are able to focus our attention and improvement actions on activities upstream of and precursor to manufacturing and assembly operations, the more effective we will be in effecting real change (improvement) in the manufacturing and assembly process which is the major focus of any quality improvement effort or process. We will therefore discuss two major elements of a Cost of Quality system: (a) the Taguchi Loss function as a measure of the quality of the (off-line) product and process engineering design and development effort; and (b) the "Conventional" cost of quality reporting systems as a measure of the on-line (production) effort. In the latter case the "Eddy Current Analysis" provides a useful adjunct to the Quality Performance Measurement System. In all cases our purpose is to:

A. Evaluate the status or and capability for implementing a total cost of Quality Reporting System and the Uses (Advantages and Cost) or such a system, and

B. Then to define the actions needed for efficient and effective implementation.

Conventional Cost of Quality

Consider the Conventional Cost of Quality System first.

While there are many ways to construct a Cost of Quality report, the easiest way to visualize the elements of the report is to picture an operation in which nothing has gone wrong and nothing can go wrong. Parts always conform to requirements, equipment does not break down, etc. Nonetheless there are still costs--referred to as standard costs of goods manufactured, or standard cost of operations, including, for example the cost of the land, physical plant, and equipment, labor, utilities, direct (and indirect) materials, etc.

In the "real world" because things can go wrong certain costs accrue above and beyond the standard cost of operations for PREVENTION and APPRAISAL activities. In spite of our best efforts to prevent failure (through such activities as preventive maintenance, design review, training, etc.) and to appraise our ongoing efforts in order to obtain a real-time, authentic signal as close to the point where (time when) the actual deviation from expectation (or requirements) occurred (through such activities as internal and supplier in-process inspection, etc.) things still do go wrong.

This results in FAILURE costs (e.g. rework, scrap, warranty, excess transportation, sorting, etc.)

The difference between total cost and standard cost--the costs of PREVENTION plus APPRAISAL plus FAILURE--is what is usually referred to as the "Cost of Quality."

Through system design we endeavor to minimize failure costs by doing the right things right the first time and, for those inevitable failures that do occur, by catching failures as close to the point of creation as possible. This can be accomplished, for example, by giving the operator real-time control information (e.g., via CRT at the work station) integrating control into the process, providing tools for evaluation and control, etc.

Detailed discussions of the measurement of the Cost of Quality are available (reference 4) and need not be repeated here. Similarly detailed discussions of system implementation and examples of several reports (including forms and instructions for their use) may be found in references.

One can visualize the Cost of Quality Report as divided into two major sections with actions or assets (prevention and appraisal/assessment/detection) on the lift and results or liabilities (failure) on the right. Typically, the cost of prevention plus appraisal plus failure is of the order of 25-35% of the standard cost of operations with failure costs accounting for 80-90% of this total. In addition, as a result of management's general preference for positive variances some failure costs are often hidden or buried in the standards.

One a Cost of Quality System is in place and at least some of these deficiencies corrected (e.g., the standards "purified" or "un-buried" so that we have a meaningful datum against which to measure) the Report can be used as an action document for:

A. Attention--To highlight the real/actual magnitude and location of the cost of un-quality. A number of the order of one third of the standard cost of goods manufactured should be large enough to get attention and action. And while, contrary to popular opinion, quality is NOT free--it is the best investment I know!

B. Prioritization--In building up to cost of quality we will identify the major contributors and identify causes. This will enable us to identify major contributors so that we attack those areas with the greatest potential/ probability for improvement (short or long term)--those with optimum return on investment.

C. Justification--Given the identification of potential areas for corrective action (item B above) and a knowledge of the cost/resources/investment required we can readily determine whether we meet the hurdles, payback, ROI or other financial investment requirements to justify/prioritize the investment decision.

D. Recognition--As progress is made (i.e., actions taken which reduce the cost of quality) this easy-to-follow indication of progress can be used to monitor, recognize and/or reward progress. This could include allocation of some of the savings for retraining, funding of additional investment for addition (other) improvements, group incentives, profit sharing, etc. (depending on managements' desires).

E. Corrective Action--The need for system improvements, reorganization, education and/or training, etc. may be indicated as a result of A-E above.

In constructing a Cost of Quality Report for your organization, choose accounts most appropriate for your operations, but remember that the simpler and easier this report is to generate, the more likely it is that it will be maintained and used. Therefore, it is useful to start out with accounts that already exist in your ledger.

As previously mentioned, in a typical "old line" manufacturing operation, it is not uncommon for the total cost of quality (prevention + detection + failure) to be as high as 25-35% of the standard cost of goods manufactured, with the cost of failure accounting for 80-90% of this total (AWARENESS). In a well run plant, the ratio of the total cost of quality to the standard cost of goods manufactured should be of the order of 5%.

Initially, as one begins to pay attention to "quality", the total cost of quality should be expected to increase followed by a shift in costs from external to internal failure (with slightly greater decreases in external costs since we are shipping fewer defectives). Finally, as the effects of our preventive measures begin to be observed, the overall cost of quality will decrease. Very early in this process (even as the costs are increasing) other measures of failure (e.g. customer complaints and rejections, late and/or erratic delivery schedules, Eddy Currents) should show significant improvements.

Typically, after an initial investment of the order of 10-20% of the cost of failure, this financial foundation for decision making shows that for most companies, generally, with concerted, well directed effort, the cost of failure can be reduced by about 50% every 1-3 years. If the cost of failure is $1 million then

Cost of Failure	Savings	Cumulative Savings	Year
$1,000,000.			0
$ 500,000.	$500,000.	$500,000.	2
$ 250,000.	$250,000.	$750,000.	4
$ 125,000.	$125,000.	$875,000.	6
$ 62,500.	$ 62,500.	$937,500.	8
$ 31,250.	$ 31,250.	$968.750.	10

32

This would seem to indicate that quality improvement (as measured by cost improvement) is about a 10 year "trip" but with significant improvement achievable during the first several years. In fact, 75% improvement is possible in as little as 4 years. And whether or not we accept the "10 year trip" estimate we can at least agree that it is a long term, ongoing process.

In this context we must recognize that, to maintain/sustain progress requires continuous, ongoing, structured improvement.

Figure 2, schematically depicts, structured improvement requires some investment (a b,c d,e f, etc.) followed by a relatively rapid but limited improvement (b c,d e, etc.). Generally the incremental investments (costs to implement improvements) are small and pay back periods short. We generally expect to save twice the cost within 6-9 months.

Thus, the process should be self funding after infusion of some seed money usually of the order of the ongoing costs of Prevention + appraisal (circa 20% of the cost of failure) for improvement projects plus 10% for Education and training.

"Eddy Current" Analysis

The "Eddy Current" Analysis was developed as a means of estimating the COQ from available Shop-floor information. Using summary (shop floor) information on test results--summary information on first run pass/fail statistics for a particular product at particular points in the process-together with production volumes for the same time frame, it is possible to determine the number of units which were cycled back through some part of the process and to estimate the number of man-hours of efforts required for these cycles.

These can readily be handled on a personal computer in spread sheet form, summarized schematically, and presented in control chart/action document form.

Eddy Currents thus obtained (fig. 3) are then used in conjunction with estimates of the rework, repair and retesting effort required (at the applicable labor hour rate) to estimate the costs associated with these Eddy Currents. Finally, a brief list of major contributors to the Eddy Current can be included (as a start to the Pareto analysis).

The average times for rework, repair and retesting are usually readily obtained from existing data based supplemented by discussions with production supervision and management. If, in production, a unit needs to be reopened for rework and/or repair it must be re-cleaned and/or re-tested (even if it passed the cleanliness test the first time through) this additional secondary rework and re-testing time must be included in the ECA.

On-Line Vs. Off-Line Quality Improvement Efforts

In the context of measure we need to develop short term (annual) quality goals and objectives in support of long-term, senior management policy/goals/objectives. This will assure that the effort is continuous and ongoing and that the strategic issues are supported by tactical actions. Coupling these actions with Statistical Problem Solving discipline, cross functional teams, open communication trust and confidence, and long term commitment, should enable us to make significant permanent in-roads into, "Quality Problems" by attacking and correcting the root causes of the problems--by treating system rather than symptom. In addition to on-line efforts, this requires quality improvement efforts in support of manufacturing and assembly operations (i.e. off-line or up-stream efforts).

Loss Function Approach

In this context, (wherein shared goals and objectives among the various functions are mandatory), I want to discuss an approach to long-term focus for the engineering design, process development and purchasing functions. New product design efforts may be grouped into 3 major categories or stages (per Dr. G. Taguchi, reference 6):

1. System design,

2. Parameter design,

3. Tolerance design.

1. SYSTEM DESIGN--This is the prototype or development stage. Generally in the U.S. this is where the major engineering effort is made. The focus here is to design a product or process which functions as intended (i.e., "at" the nominal specification value) when new/perfectly "tuned." No serious effort is made to assure that the product or process is ROBUST--that the product or process will continue to function "at" nominal with age or under normal use where noise (or uncontrolled external variation) is present. In this phase of design we are concerned about nominal values to the exclusion of issues related to common and special causes of variation.

2. PARAMETER DESIGN--In this second phase of the design and development process, we attempt to achieve parameter optimization--to reduce variation without cost increase. The goal is to identify the best set of nominal values of parameters so that external variations ("noise") cause as little variation as possible in the product or process. We design the product or process so that the parameters are as robust as possible. This phase is found to be exceptionally cost effective. The results can usually be achieved with no added cost. Often there are attendant cost reductions.

3. TOLERANCE DESIGN--In this third and last step in the design process, we attempt to reduce or remove causes of variation, and achieve quality improvement by selectively upgrading parameters and/or components. Engineering and manufacturing tolerances can then be specified. To accomplish these latter ends cost effectively, and simultaneously provide the basis for measuring the design effort, Taguchi recommends (reference 6) that we recognize loss (or cost) due to variation (fig. 4b). Rather than the conventional approach wherein cost is only assumed to occur when we exceed design specification limits (fig. 4a), we should adopt the more realistic approach that there is loss (cost) due to all variation; that this loss is a minimum (i.e., zero) when the parameter or characteristic (Y) is at nominal (i.e., when Y = m) and increases at an increasing rate as we deviate from nominal (fig. 4b). How this would work can be most easily visualized by example.

Suppose a power supply is used to regulate/control a motor and the customer's tolerance was 10±2. Once the output is outside this range (8-12 VDC) the customer has to repair the entire assembly at a cost of $100.00 on average for labor and materials. Suppose further that the output voltage can be calibrated during assembly by changing one resistor at a total cost of $10.00. When would we spend the $10.00 for in-process calibration (i.e., what should the production--assembly--tolerance be)?

Following Taguchi's suggestion, (reference 6), we use:

$L(Y) = k(Y-m)^2$ for the loss function then, from the information on external repair costs we can solve for k.

$$L(Y) = k(Y-m)^2$$

$$\$100 = k(8-10)^2 = k(12-10)^2$$

$$k = 25$$

and from the information on in process calibration costs we can obtain the production tolerance limits Y_{pt1}.

$$L(Y) = \$10.00 = 25(Y_{pt1}-10)^2$$

$$Y = 10 \pm 10/25$$

$$Y_{pt1} = 10 \pm 0.63$$

The production tolerance is therefore 9.37-10.63. In otherwords, we would calibrate the power supply if the output voltage is outside of 10 ± 0.163 VDC and not calibrate otherwise.

The above calculation, for a single piece can easily be generalized to many units simply by taking the average of $(Y-m)^2$. For n pieces

$$L(Y) = k * \frac{1}{n} \quad (Y_i - m)^2$$

$$= k * \frac{1}{n} \quad (Y_i - Y)^2 + (y - m)^2$$

$$= k \quad {}^2 + (y - m)^2$$

To reduce L(Y) we have to

1. reduce variation around the average (i.e. reduce ( ) and/or

2. reduce the difference between the process average (Y) and nominal (m): (i.e., adjust the average to the nominal so that $(Y - m)^2$ is negligible).

$L(Y) = k * {}^2 + (Y - m)^2$ and hence the monthly or annual difference and thereby evaluate the relative merits/efficacy of competing designs, competing processes, competing vendors. And by corrolary we can similarly evaluation product and process design and purchasing activities.

Epilogue

Quality is the attitude that infiltrates all that a company does whether by direction from above (lead!), with support from above (follow!), or with sufferance from above (get out of the way!). In point of fact, a combination of all three (direction or focus plus support and sufferance) is the most efficacious combination and one that will guarantee success. The key issue is not only whether management knows and understands their jobs, but also whether a particular management style is used. The main ingredients required if an organization is to effectively produce quality products are: (1) philosophical focus; (2) specific goals developed within the various organizations in support of the philosophical focus; (3) knowledge not only on the part of managers, but all participants within the organization; (4) an open and

participative form of management; (5) employee involvement; and (6) positive reinforcement in order to create new behaviors. AT some level in the organization, there
must be a leader. However, if the focus is understood and the commitment exists,
then leaders will be found at all levels within the organization. In this context,
control for quality is achieved through goal setting supported by an information
system which enables all members of the team to chart progress. At the management
level, the Product Review and Evaluation Program (PREP), the Cost of Quality Report
(COQ), and the Numerical Quality Index of Customer Performance or the Customer Complaint and Rejection Report (C2R2) provide an outstanding combination of strategic
control coupled with tactical freedom of action.

It is strongly recommended that all organizations implement a Quality Performance
Measurement System. This system should include, at a minimum, the following elements:

1. The cost of Quality Report as the strategic senior management action document.

2. The Eddy Current Analysis (ECA) as the tactical operations management action
   documents.

3. The loss function analysis as the tactical staff management (product design,
   process development, procurement) action documents.

4. The Product/Process Review and Evaluation Program (PREP) as the check-
   list/discipline for the entire new product design and development (technology
   transfer) process, and

5. The Customer Complaint and Rejection Report (C2R2) as a measure of external
   performance (customer satisfaction).

It is not that we need to capture more information necessarily, we merely need to
capture the information in different ways. This is not necessarily more time consuming. In fact, it may be less time consuming. In fact, the cost to implement COQ
need not be large. Initially, it could be as simple as introducing a coding system
that stratifies variances into appropriate categories (prevention, appraisal, failure, etc.).

By making "Un-Quality" associated costs visible through the Cost of Quality Reporting
System (CQRS) action to drive down cost will be forthcoming. Actual results are
obviously, dependent on efficacy of other programs since the measure discussed in
this section to not themselves cause improvement. Nonetheless, the CQRS sets the
stage by prioritizing problems and providing the information needed to justify their
solution. In short, "Quality is not free--but it is the best investment known!"

Early or	Random or	Design - Life
Premature	Steady State	Failure
Failure Region	Failure Region	Region

                                                                        TIME

FIGURE 1
   BATHTUB CURVE OF FAILURE

36

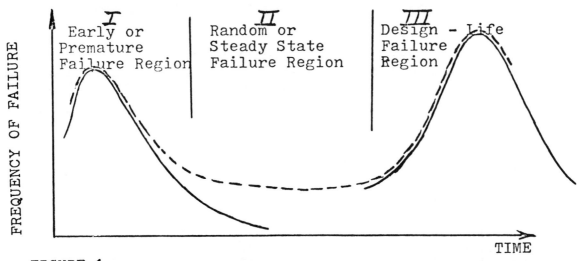

FIGURE 1
BATHTUB CURVE OF FAILURE

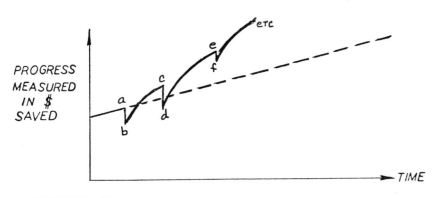

FIGURE 2
THE PROCESS OF CONTINUOUS
STRUCTURED IMPROVEMENT

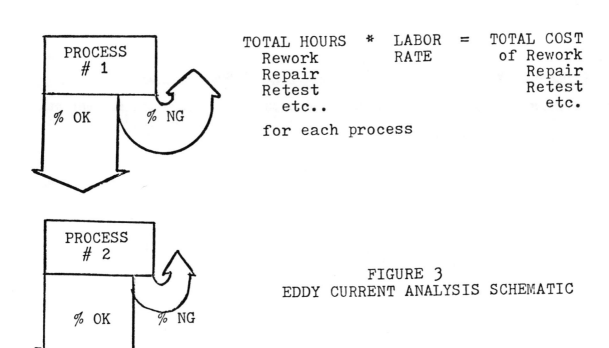

TOTAL HOURS * LABOR = TOTAL COST
  Rework   RATE   of Rework
  Repair   Repair
  Retest   Retest
   etc..   etc.

for each process

FIGURE 3
EDDY CURRENT ANALYSIS SCHEMATIC

LOSS BY DISPERSION

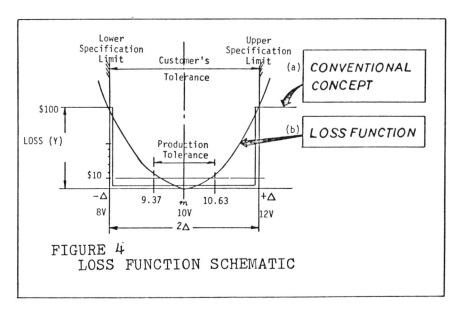

FIGURE 4
LOSS FUNCTION SCHEMATIC

APPENDIX A

PRODUCT REVIEW AND EVALUATION PROGRAM (PREP)

In order to assure the successful launch of new products, it is useful to formalize the elements involved in the launch process. Exhibit A lists the major elements of the launch process for a metal stamping supplier company in the automotive industry and the activity with primary responsibility for each element. Overall coordination and tracking of the program is the responsibility of Engineering.

Exhibit B shows the summary sheet which will be used to record the current status of each element for each part. Status reviews will be scheduled at the direction of the Plant Manager or the Engineering Manager.

In addition to instituting this program on all new parts, a modification of the program should be implemented on selected current parts. The modification of the PREP for current parts is the omission of the starred items in Exhibit A.

Going through this formal analysis on existing parts will aid in the identification of problems for resolution. Applying this method to new parts should greatly reduce the number of problems experienced when these parts get into production.

CHECK-LIST--Exhibit A

ITEM	RESPONSIBILITY
1. Determine the latest engineering level required by customer.	Material Control
2. Obtain customer print.	Engineering
3. Obtain customer specification referenced on print.	Engineering
4. Prepare company print and verify with customer print.	Engineering
5. Designate essential and major characteristics.	Engineering
6. Issue company print.	Engineering
7. Prepare and issue material specifications.	Engineering
8. Prepare and issue routing.	Engineering
9. Prepare and issue process sheets.	Engineering
*10. Place orders for design and build of tooling.	Engineering
11. Prepare purchase orders for materials and sub-contracted services.	Material Control
12. Prepare inspection instructions for each inspection point.	Quality Control
13. Determine gaging requirements and prepare design and build requests.	Quality Control
14. Place orders for gaging design and build.	Engineering
15. Certify gaging, establish recert frequency and prepare recert procedures.	Quality Control
16. Certify tooling.	Quality Control
17. Prepare packaging and labeling specifications.	Material Control
*18. Produce engineering prototypes.	Engineering
*19. Inspect prototypes.	Quality Control
20. Product initial samples.	Material Control
21. Inspect initial samples.	Quality Control
*22. Produce first production lot.	Material Control
*23. Inspect first production samples.	Quality Control
24. Perform process capability studies on essential characteristics.	Quality Control

Exhibit B

PRODUCT REVIEW AND EVALUATION PROGRAM

OEM PART NUMBER _____   CHANGE LEVEL _____

CUSTOMER _____   TOOLING ORDER RECEIVED _____

DESCRIPTION _____   ISIR REQUIRED _____

PART NUMBER _____   FPS REQUIRED _____

1) Determine Engr. Level	2) Obtain Customer Print	3) Obtain Customer Specs.	4) Prepare D.A.B. Print
Init. Curr. Act. Prom. Prom. Comp.	Init. Curr. Act. Prom. Prom. Comp.	Init. Curr. Act. Prom. Prom. Comp.	Init. Curr. Act. Prom. Prom. Comp.

5) Designate Characteristics	6) Issue D.A.B. Print	7) Issue Material Specifications	8) Issue Routing
Init. Curr. Act. Prom. Prom. Comp.	Init. Curr. Act. Prom. Prom. Comp.	Init. Curr. Act. Prom. Prom. Comp.	Init. Curr. Act. Prom. Prom. Comp.

9) Issue Process Sheets	10) Tooling Design & Build	11) Prepare Purch. Orders	12) Prepare Insp. Instructions
Init. Curr. Act. Prom. Prom. Comp.	Init. Curr. Act. Prom. Prom. Comp.	Init. Curr. Act. Prom. Prom. Comp.	Init. Curr. Act. Prom. Prom. Comp.

13) Determine Gaging Reqmnts.	14) Gaging Design & Build	15) Certify Gages Estab. Recert.	16) Certify Tooling
Init. Curr. Act. Prom. Prom. Comp.	Init. Curr. Act. Prom. Prom. Comp.	Init. Curr. Act. Prom. Prom. Comp.	Init. Curr. Act. Prom. Prom. Comp.

17) Prepare Packaging Spec.	18) Produce Prototypes	19) Inspect Prototypes	20) Produce ISIR Samples
Init. Curr. Act. Prom. Prom. Comp.	Init. Curr. Act. Prom. Prom. Comp.	Init. Curr. Act. Prom. Prom. Comp.	Init. Curr. Act. Prom. Prom. Comp.

21) Inspect ISIR Samples	22) Product FPS Lot	23) Inspect FPS Samples	24) Demonstrate Capability
Init. Curr. Act. Prom. Prom. Comp.	Init. Curr. Act. Prom. Prom. Comp.	Init. Curr. Act. Prom. Prom. Comp.	Init. Curr. Act. Prom. Prom. Comp.

APPENDIX B

COST OF QUALITY REPORT

In order to track the cost of quality it is necessary to establish accounts in each of the major cost categories: Prevention, Appraisal, Internal and External Failure. Then specific rules with respect to use/meaning of each account need to be established. The cost of the same physical activity will be placed into different accounts depending on the REASON for the activity. For example, testing to determine whether a product meets design intent prior to committing to production is a prevention cost; similar testing in process to assure that the process is producing parts "to print" is an appraisal cost; and the same testing activity to sort good from bad in a rejected lot is a failure cost. Furthermore, if the rejected lot is found in our (or our supplier's) plant it is an internal failure cost. The same lot rejected by our customer would make the testing an external failure cost.

Exhibit A lists some examples of account categories. However, without detailed understanding of the intent/meaning of each of the accounts, difficulties will be encountered. In actual use in a multi-plant manufacturing company, after a year of allowing each plant to use their own version of Exhibit A, it appeared that we would gain a better understanding of the costs associated with quality and provide uniformity in plant-to plant reporting by standardizing on Exhibit B.

In many cases, the comptrollers had to have new account numbers assigned by Corporate Staff in order to track the various items. To protect salary confidentiality, we combined the items shown as 1-A, B, C, & D (in Exhibit B of the attachment) into only line item. The attached memo explains and interprets the various line items in Exhibit B.

Exhibit A

COST OF QUALITY (BY PART #)

Prevention

A. Quality Control Manager
B. Quality Control Engineering
C. Calibration Program
D. Metrology Supervisor
E. Set Up
F. Design Reviews
G. Product Qualification
H. Drawing Checking
I. Engineering Quality Orientation
J. Make Certain Programs
K. Supplier Evaluations
L. Supplier Quality Seminars
M. Specification Review
N. Process Capability Studies
O. Tool Control
P. Operator Training
Q. Quality Orientation
R. Acceptance Planning
S. Zero Defects Program
T. Quality Audits
U. Preventative Maintenance
V. Process Improvement Programs
   Aimed at Quality Improvement

COST OF UNQUALITY (BY PART #)

Failure

A. Sort (man-hours)
B. Rework (man-hours)
C. Scrap (the cost already in the part) by department
D. Returns (sort, scrap, rework plus twice the shipping cost)
E. Customer Charges (sort, scrap, rework, etc.
F. Extra Operations/Rework
G. Downtime Due to Quality Problems
H. Consumer Affairs
I. Redesign
J. Engineering Change Order
K. Purchasing Change Order
L. Corrective Action Costs
M. Warranty Costs
N. Service
O. Product Liability
P. "Emergency"/Non-Planned Maintenance
   1. Downtime (man-hours and overhead)
   2. Repairs (man-hours)
   3. Hardware ($)

Detection/Appraisal

A. Inspection Supervisors
B. Inspectors
C. Gaging
D. Prototype Inspection & Test
E. Production Specification Conformance Analysis
F. Supplier Surveillance
G. Receiving Inspection & Test
H. Product Acceptance
I. Process Control Acceptance
J. Packaging Inspection
K. Status Measurement & Reporting

NOTE: We may want to separate INPROCESS AND FINAL INSPECTION costs in each category.

D.A.B. Industries, Inc.
Quality & Reliability
Engineering
June 23, 1980

42

INTER-COMPANY MEMO FROM: W. R. Stackhouse, Corporate Director
                          Quality and Reliability Engineering

SUBJECT: Cost of Quality Reporting Practices

About a year ago when the original list of items to be considered in the "Cost of Quality" was published (Exhibit A), we had just determined that tracking the costs associated with "quality" and "unquality" seemed like a good idea and attempted to list all those items which appeared to be associated with "quality". The plants were not asked to change any of their accounting procedures, but to use their current methods to try and put a cost figure on as many of the listed items as they could.

After a year of viewing the results of this attempt and solidifying our thoughts on the intent and usefulness of this type of report, I feel it is time to formalize the structure and actually account for the items we feel are beneficial to track.

Exhibit B shows a listing of the line items which I feel should be used in compiling a report on the quality costs associated with the three phases of Prevention, Detection and Failure. A brief explanation of some of those items is as follows:

1-A, B, C, & D

The QC Manager, QC Engineers, Metrologists and Clerk are engaged primarily in administering the quality program, determining capabilities, designing process controls, certifying gages and other quality system related activities and as such, their salaries and fringes should be charged to Prevention.

1-E & F

In support of the metrology Program, certain gages or measuring devices must be outsourced for calibration or recertification. These charges are preventive in nature as are the costs of gages and measuring devices used in the Metrology Lab for in-house calibrations and layouts.

1-G

Periodic layouts performed on the First and Last piece of a production run are preventive in that they are intended to alert the tool room to die and tool fixes required prior to subsequent runs. The layout of samples also falls into this category (not the sorting of 500 pieces to select 10 good ones to show the customer, but the layout of samples to provide manufacturing and tool engineering with information). The time that an auditor spends doing this type of work (vs. audit inspection or acceptance sampling) should be changed to Prevention.

11-A & B

The salary and fringes of the supervisor of the audit inspectors is primarily Detection as is the time spent by the auditors doing audit or acceptance sampling (not 100% bench inspection).

11-C & D

Gaging of instrumentation tailor-made for checking specific parts and general measurement instruments used by the auditors (micrometers, calipers, etc.) and the repairs to these items should be charged to Detection.

111-A

In-house Failure is the cost incurred as the result of us finding a problem ourselves before the parts are shipped.

111-A1, 2 & 3

Labor costs associated with sorting and reworking rejected material and the standard cost of scrapped material is a cost of Failure.

111-A4 & 5

When equipment is shut down for quality reasons, the labor spent on resetup as well as the downtime incurred by the operators (if they are idle during resetup and not assigned other work) is a cost of failure.

111-B

Return Failure is the cost incurred as a result of problems begin found by the customer and the parts returned.

111-B1, 2 & 3

Same as 111-A1, 2 & 3.

111-B4

The freight charge for returned material is part of the cost of Failure and should be entered as double the charge (once for the return--once for replacing the shipment).

111-C1, 2 & 3

In some instances, the customer may find a problem and elect to sort or rework the material himself. The bill from the customer or the credit given his account should be charged as a cost of Failure. The standard cost of material scrapped by the customer should also be charged to Failure.

111-D

The Sale of Scrap is a minor offset to the cost of Failure.

Some ratios which would be meaningful and could be used for comparison purposes are:

Cost of Quality Ratio

(1 + 11 + 111)/Standard Cost of Manufacturing

Cost of Failure Ratio

(111)/Standard Cost of Manufacturing

In-House Failure Ratio

(111A)/(111A + 111B + 111C)

Scrap Ratio

$$(111A3 + 111B3 + 111C3 - 111D)/111$$

In order to track quality costs in this manner, accounts would have to be set up in each plant for each of these lines items and ground-rules established for charging to these accounts.

COST OF QUALITY REPORT--Exhibit B

I. <u>Prevention</u>

    A. QC Manager

    B. QC Engineers

    C. Metrologists

    D. QC Clerk

    E. Metrology Program

    F. Metrology Equipment

    G. Layout Inspection

II. <u>Detection</u>

    A. Inspection Supervision

    B. Audit Inspection

    C. Dedicated Gaging

    D. General Gaging

III½ <u>Failure</u>

    A. <u>In-House</u>

        1. Sort
        2. Rework
        3. Scrap
        4. Resetup
        5. Downtime

    B. <u>Returns</u>

        1. Sort
        2. Rework
        3. Scrap
        4. Freight

    C. <u>Customer Charges</u>
        1. Sort
        2. Rework
        3. Scrap

    D. <u>Sale of Scrap</u>

REFERENCES/BIBLIOGRAPHY

1. AARON, HOWARD B., "Lead, Follow, or Get Out of the Way!", Transaction of the 36th Annual Quality Congress of the American Society for Quality Control, May, 1982, page 21.

2. AARON, HOWARD B., "Control for Quality: Not Quality Control.", Automach Australia '84 Conference Proceedings, May 23-25, 1984, Sydney Australia, pp 4-17 to 4-38, Society of Manufacturing Engineers, Dearborn, Michigan, U.S.a., and

   AARON, HOWARD B., "Control for Quality: Not Quality Control--Management and Motivation." To be published in the 1986 American Society for Quality Control Congress Transactions, Anaheim, California, U.S.A., May 19-21, 1986.

3. BAJARIA, HANS J., "Statistical Process Control (SPC) and Automation." To be published in the 1986 American Society for Quality Control Congress Transactions, Anaheim, California, U.S.A., May 19-21, 1986.

4. SCHMIDT, J.W. and JACKSON, J.F., "Measuring the Cost of Product Quality.", SAE paper # 820209 contained in SAE publication SP-512 "Effective Quality Cost Analysis System for Increased Profit and Productivity," International Congress and Exposition, Detroit, Michigan, U.S.A., February 22-26, 1982.

5. AARON, HOWARD, B., BAJARIA, HANS J., and KRAG, WILLIAM B., to be published.

6. TAGUCHI, GENICHI and WU, YUIN, "An Introduction to Off-line Quality Control," Central Japan Quality Control Association, February, 1985, p. 7 et seq., and Quality Engineering--Product and Process Design Optimization by Prof. Yuin Wu and Dr. Willie Hobbs Moore, American Supplier Institute Inc., Six Parklane Boulevard, Dearborn, Michigan, U.S.A. 48126, September, 1985.

7. BAJARIA, HANS J., "'Quality Circles'--Will They Work for Your Organization?", Automotive Division Newsletter of the American Society for Quality Control, May 1981.

8. SHAW, F., "The Design Approach to Reliability", SAE paper 741109, 1974.

9. SULLIVAN, L., Company-Wide Quality Control for Automotive Suppliers, Ford Motor Company, Dearborn, Michigan, U.S.A., 1985.

10. BERRY, BRYAN H., "Preventing Defects Through Statistical Quality Control", Iron Age, October 1981, page 59.

11. Business Week, March 12, 1972, Special Report, page 32B.

12. CALLAHAN, JOSEPH M., "The Deming Era--New U.S. Industrial Revolution?", Automotive Industries, December 1981, page 23.

13. CROSBY, PHILIP, Quality Magazine, August 1980, page Q15.

14. DEMING, W. EDWARDS, "Dr. Deming's Cure for U.S. Management", Ward's Auto World, November 1981.

15. DRUCKER, PETER F., "What We Can Learn From Japanese Management", <u>Harvard Business Review</u>, March-April 1971, page 110.

16. HANCOCK, WALTON M., and PLONKA, FRANCIS E., "The Development of Quality Information Systems in Automotive Assembly", SAE paper 770414, 1977.

17. KONZ, STEPHAN, "Quality Circles--An Annotated Bibliography", <u>Quality Progress</u>, April 1981, page 30.

18. MULLER, HERBERT J., <u>The Uses of the Past--Profiles of Former Societies</u>, New York: Oxford University Press, 1957, page 5 et seq.

19. OUCHI, WILLIAM, <u>Theory Z: How American Business Can Meet the Japanese Challenge</u>, Reading, Massachusetts: Addison-Wesley Publishing Co., Chapter 2, page 39 et seq.

20. PETERS, GEORGE A., "The Decade Ahead", <u>Quality Progress</u>, September 1974, pp. 22-23.

21. Ward's Auto World, December 1981, page 25.

22. AGEE, WILLIAM, <u>The Detroit Engineer</u>, January 1982, page 27.

23. AARON, HOWARD B., "Suppliers: The Hourly Workers of the Auto Industry.", Automach Australia '85 Conference Proceedings, July 2-5, 1985, Melbourne, Australia, pp 7-1 to 7-11, Society of Manufacturing Engineers, Dearborn, Michigan, U.S.A. 48121.

If additional information is requested please feel free to contact us.

Fabricators and Manufacturers Association, Intl'
5411 E. State Street
Rockford, Il 61108
815/399-8700

Presented at the SME Tube Fabricating Technology
Conference, March 1985

Statistical Process Control: What It Is and Why We Need It

David A. Marker and David R. Morganstein
Westat, Incorporated

This paper discusses the importance of upper management's role in the implementation of statistical process control. Why are so many companies concerned with statistical thinking? What can it do for my company?

We first describe what we mean by statistical thinking, and what is involved in measuring the cost of quality. Then, we compare the traditional view of quality with the newer "Japanese" view. Finally, we cover some basic statistical definitions, elements of statistical training, and the need for standards.

Many companies have turned to statistical techniques in an attempt to avoid going out of business. Others have become involved in response to pressure from customers such as Ford and GM. Still others have noted that while the West has not stood still in respect to quality, over the last 30 years, the reputation of Japanese goods has gone from "use it today before it breaks" to "the very best quality for the dollar." What has happened? What do they know that we don't?

The effects of this quality differential can be seen quite clearly in the following table comparing the inventory costs of Japanese and American automobile makers. The Japanese typically have a 10-part per million (ppm) on-line rejection rate, compared to the 1 to 4 percent of American car companies. Both make cars with approximately 10,000 parts. The low Japanese defective rate results in an extra billion dollars of inventory needed to produce American cars!

Table 1. Comparison of Japanese and American car manufacturing

Japan	U.S.
10-ppm reject rate on-line or 0.001 percent defective	1 to 4 percent defective
10,000 parts per car	10,000 parts per car
1 part per 10 cars replaced on-line	100 to 400 parts per car replaced on-line
1 to 6 hour spare parts inventory	1 to 3 month spare parts inventory
$0 in inventory	$1 Billion in inventory

In response to these differentials, many western companies have started to examine the Japanese methods and to realize that they too must incorporate statistical thinking into their decision-making if they are to compete in the international market place. Both Ford and GM now require of their suppliers an ongoing plan demonstrating management's commitment to quality, including the presentation of control charts accompanying each order. This in turn has led the management of all their suppliers to try to discover the secrets of statistical process control.

The first book to discuss in detail how to implement SPC was Walter Shewhart's 1931 book, Economic Control of Quality of Manufactured Product. The role of management in SPC is described in W. Edwards Deming's Quality, Productivity, and Competitive Position. Many of the ideas included in this paper are also discussed in these books.

John Johnston, the former Director of Research at U.S. Steel, said:

> "The possibility of improving the economy of steel to the customer is therefore largely a matter of improving its uniformity of quality, of fitting steel better for each of the multifarious uses, rather than of any direct lessening of its cost of production."

This would not be particularly noteworthy except that Mr. Johnston made this comment during the 1930's! The ideas of SPC are not new; it is only the growing recognition of their applicability that is new.

COST OF QUALITY

Many companies have begun to compute their cost of poor quality. There are many visible costs such as downgrades, rework, and scrap, but there are also a great many less visible costs such as inspectors and down time. Still, other costs are impossible to accurately measure. For example, how much did you lose last year due to business lost or never acquired because of a bad reputation? Feigenbaum, in a recent issue of Quality Progress estimated that while a happy customer tells 8 others, an unhappy one tells 22 others!

Many companies in the West have estimated their cost of quality in the 10 to 15 percent range, with some up near 20 percent. (The Japanese typically consume only two to four percent, most of which covers training and data collection.) Thus, a company with gross sales of $1 billion had a cost of quality of between $100 and $200 million. What could you do with that money? Are there any capital investments that you think you need that might be funded through quality savings? Maybe immersion ultrasonic facilities for $150,000 or continuous annealing lines. Do you have an analysis that clearly demonstrates whether your current equipment is being operated most efficiently?

It should be clear to anyone that the potential savings far exceed any increased costs that may arise in the implementation of statistical process control. Figure 1 demonstrates this point graphically. As you put more of your efforts into an SPC program (moving to the right in the diagram), the increased quality will result in greater and greater net savings.

FIGURE 1.

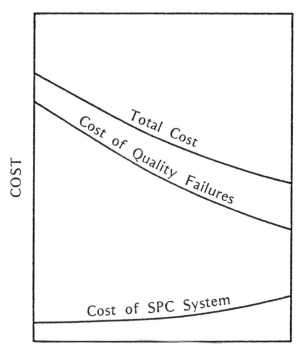

QUALITY

RESPONSIBILITY FOR QUALITY: OLD AND NEW PERSPECTIVES

We have already mentioned the need to modernize the equipment in our plants. This requires large amounts of capital. How do we know whether the gains to be anticipated through improved technology will be worth the expense? Which investments will have the higher return?

To answer such questions it is necessary to know what the present production system is capable of producing. It is also necessary to be able to measure the effect of a change in the one process that is to be replaced. Is it possible to measure such changes in your company? To do so, you must be taking measurements at every major step in the production process. Only then can you measure the effect of a change to one process. It is for keeping track of just such measurements that a control chart is designed. The chart consists of plots of average values and ranges from small samples taken over a period of time. Using the chart, it is possible to understand how much variability is built into the system, and how much of it is a result of special, unique causes.

Recognizing that much of the variability is built into the processes of production, and therefore should not be blamed on the hourly worker (indeed Dr. Deming estimates that over 85 percent of all variation is a responsibility of management), implies that quality is everyone's responsibility, from the chief executive officer, all the way to the hourly worker. Thus, statistical tools must be taught to everyone in the company and must be used by everyone if it is to have a chance to succeed.

Where does this variability in the production process come from? There are many possible sources. There may be multiple streams in a casting or rolling operation that are not producing equivalent products. Are the temperatures and pressures the same in each stream? Even if there is only one machine, there are still many possible sources of variation. Is preventive maintenance performed to ensure the consistent reliability of the operation? Do the operators on each turn run the equipment in the same way? Since they usually have no way of knowing whether the process is behaving properly at the beginning of their shift, they typically begin the day by "tweaking" the process until it meets their preferred operating conditions. If the system was already under control, all they have done by tweaking is to increase the variability of the downstream product. The result of all this variability shows up as different billet weights, tube diameters, eccentricities, etc.

Figure 2 demonstrates the tradition view of a production process, where station 1 might be a caster, station 2 extrusion presses, and stations 3 and 4 finishing processes. Alternatively, these could refer to four successive procedures in the casting process. The important point is that traditionally a product is made <u>and then</u> it is inspected for quality. What happens when the inspected product is rejected? Everyone is demoralized and tries to adjust to prevent it from happening again. How does anyone know which process needs to be changed and which should be left alone? Without data on each individual process, you cannot determine the appropriate response to prevent a reoccurrence of the problem. Similarly, without adequate data it is impossible to decide where to make major capital investments. The end result of this traditional view of quality is the continued production of scrapped tubes, coils, bars, etc.

FIGURE 2.

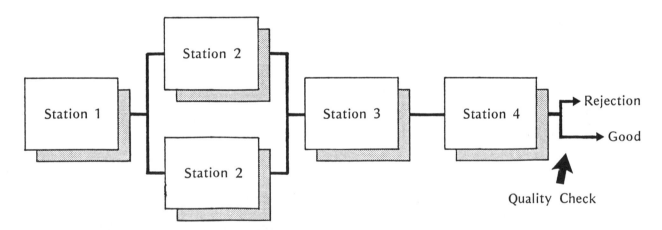

Another result of traditional quality control is a large inspection department. How many tons did your inspection department roll last month? How many inspectors does your company have? How many of your customers have to inspect your product?

Many companies have large maintenance departments to handle all of the unexpected repairs. How much does this downtime cost your company? Use of statistical procedures to monitor equipment will allow you to develop a system of preventive maintenance which will drastically reduce downtime.

We have all seen tubes with hold stickers on them resting below a sign reminding workers to "make it right the first time." If any operator has no better tool than the traditional view described above, then we are resigned to forever seeing this unfulfilled promise. Management must supply the work force with a better tool with which to do its job.

Quality cannot be left to a Quality Control Department. Instead, everyone must view quality as their own responsibility. This modern view is demonstrated in Figure 3. We have the same four stations as before, only now data is collected at each station, not just after the product is finished. The little curves by each station represent the measured variability at that step in the process.

FIGURE 3.

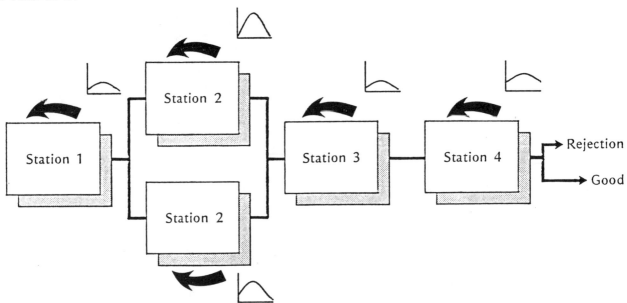

At each station people must monitor the processes that they have some control over, not the incoming materials or outgoing product. At an extrusion press, you could monitor the temperature of the billets, the quality of the lubricants, and the condition of the dies. You do not want to wait for a curved tube to be produced and then try to straighten it out. Dr. Deming compares the logic of producing a curved product and then trying to straighten it with, "You burn the toast and I'll scrape it!" It clearly costs less to produce a straight tube than to fix a curved one.

EXAMPLE

A tube producer for an automobile company was using inspection to ensure the delivery of a product that was within the specifications. The customer knew enough statistics to recognize from the distribution of diameters on received products that approximately 20 percent of all tubes being produced by the supplier were beyond the upper spec. The customer also knew enough to realize that somewhere down the line he was being charged for this 20 percent! He demanded that the producer improve his production process to eliminate the 20 percent. The supplier first realized that the hand micrometers they were using were too inconsistent to provide the necessary accuracy; they replaced them with digital mics. Then, they brought the system into statistical control by using statistical control charts on the production line. With the aid of additional statistical techniques, they were able to identify major sources of variability, eliminate them, and reduce the reject rate from 16 percent down to 4 percent. This has resulted in annual savings to the tube producer in six figures. The costs of the statistical analysis and digital micrometers were minuscule by comparison.

STATISTICAL TOOLS AND TRAINING

Figure 4 shows the monthly production of three operators of equal quality, each of whom averages 20 tons per hour. Do you consider these operators of equal performance? Very few would like to have Joe working for them. Some prefer Frank, but most want to work with Charlene. Why? What is it beyond the average product that concerns people? Joe is so unstable that you have no confidence in his production for next month. Frank is extremely stable and predictable. Charlene's consistent improvement allows us to predict better future production levels for her than the others. Variability and trends over time, along with our ability to understand their causes, are at least as important in our choice of operators as is their average production. We will only be able to understand these characteristics after learning how to use and interpret a few statistical tools.

FIGURE 4.

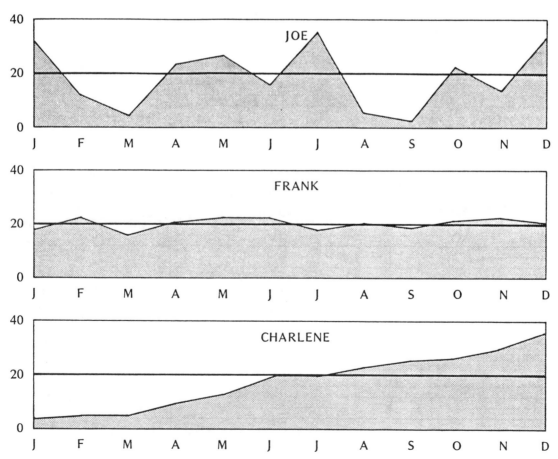

Similar confusion often arises when we try to interpret data on rejection rates. Rejections for a particular company had been averaging 8 percent and now average only 5 percent. After a period of celebration, it is noticed that rejections have suddenly reached 9 percent. How often have you heard someone say, "I thought we solved that problem six months ago!" What went wrong? Was 5 percent really statistically different from 8 percent? By this we mean that if traditionally we varied between 7 and 9 percent, then a period of only 5 percent rejects may represent a true improvement. If, traditionally, rejects were between 4 and 12 percent, then the 5 percent rate did not indicate a change in the system of production. A change in an average is only significant (worth celebrating) when it is large compared to the variability present in a stable system. How much time does your company spend responding to morning calls demanding an explanation for the rise in rejects from 8.2 to 8.4 percent? Without using statistical techniques to track the variability in the process, you cannot tell whether this rise has a specific assignable cause that you could identify, or whether it is simply due to the variability in the system.

The purpose of statistical training is to develop an objective procedure whereby each employee can understand the data produced by the processes he or she is responsible for, and can interpret them in a manner agreeable to all in the company. Training, however, is only a first step; procedures must be developed to implement the training into the day-to-day corporate operations. Only in this way can full advantage be made of the statistical techniques.

The level of training should vary with the responsibilities of each employee. All levels of management must take part in sessions designed to familiarize them with the basic statistical techniques such as measuring variability and using control charts. This is important for two reasons: management must begin to use these tools as part of its basic decision-making process, and it is an excellent way to demonstrate the level of management's commitment to SPC to other employees.

Other salaried employees need to undergo a more thorough training in statistical techniques. They must understand a variety of different types of control charts, as well as how to set up and interpret them. Some of these will become internal consultants and teachers, and they, therefore, must be aware of more advanced statistical techniques such as regression, design of experiments, and analysis of variance.

Hourly employees should at a minimum be exposed to the ideas of a control chart, how to plot and interpret points. Many companies will find these separations unnecessary and will train hourly employees similarly to salaried ones.

Some statistical concepts should be taught to all employees. The ability to draw conclusions from samples rather than examining the whole population is not new to accountants or nurses drawing blood. Its applicability to operating departments must also be made clear.

The use of histograms and other graphical techniques to summarize large data sets needs to be encouraged. No one examines all of the numbers on a computer output; a picture does a far better job of conveying the important points. Save computer outputs for detailed searches for the cause of an irregularity.

The main shortcoming of histograms is that it is hard to compare two of them in a few words. Therefore, we must employ summary statistics that allow us to compare the histograms from two or more samples. We use the mean or median to estimate the center of the distribution, and the range or standard deviation to estimate its spread.

WHAT IS QUALITY?

We conclude by reexamining our definition of quality. Traditionally, quality has been viewed as conformance to standards. Zero defects. Anyone familiar with the experience of Ford's Batavia, Ohio, transmission facility realizes that zero defects is no longer sufficient.

Figure 5 shows a sequence of five diagrams. The first demonstrates the traditional view that any product within the specifications is fine, and any beyond the specs is totally useless. In reality, our loss is more likely to follow the function shown in the second picture; there is no loss if we hit the aim perfectly. There is an increasing loss as we deviate further from the aim.

FIGURE 5.

SPECIFICATIONS

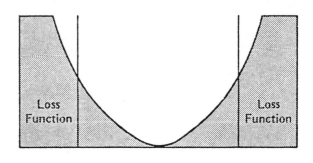

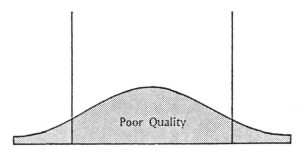

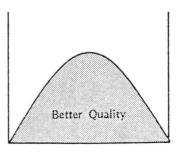

What does this imply for our definition of quality? Everyone would agree that the third diagram of Figure 5 indicates poor quality. But the above loss function implies that while the fourth diagram shows good quality (zero defects), the fifth shows even better quality! We will continue to improve the quality <u>only be continually reducing the variability</u>. This is the reason for Ford Motor Company's emphasis on *Continuing Process Improvement*, not just process control.

CONCLUSION

Quality can only be improved through the use of statistical techniques to measure and reduce the variability in the system of production. Relying on inspection to fight fires as they arise guarantees that you will continue to produce the same quality product you presently produce. SPC is a long-term philosophy, but it must begin with top-level management. Only when they start to make decisions through an understanding of statistical thinking will all employees be free to work towards improving the quality of the products they produce.

Reprinted from *Tooling and Production*, October 1982

Producing quality for the '80s

SQC in real time

Statistical quality control (SQC) offers us the techniques to bring our manufacturing processes under control and deliver the quality we will need to stay competitive in the '80s. Now minicomputers offer us the tools to make statistical calculations and plots in real time, fast enough to modify the process before a single bad part is made.

Remember when we couldn't make enough? No matter what it was we manufactured, an insatiable marketplace was there to absorb our products. We became obsessed with a "make it faster" philosophy and weren't too concerned that not all the product was right the first time. We began to plan for scrap and rework and set aside manufacturing areas specifically for these problems. Quality-control departments attempted to final-inspect quality into the product. But by then, of course, it was too late. The quality was either there or it wasn't.

We began to buy dedicated special machinery that could produce products even faster. Of course, it was difficult to redeploy these machines to make model changes, so we kept making the same products, even when the market told us it wanted something new.

We placed tremendous emphasis on increasing the speeds and feeds of machine tools. Cycle time was king. It didn't matter that the tooling wore out too fast or that the surface finish of the product was so poor that the product was almost nonfunctional. Preventive maintenance was something everyone talked about and never did. We didn't mind a little downtime now and then because when those machines ran, they really ran!

by **Michael J Dickinson**
Senior Vice President, Engineering
Hansford Data Systems Inc
Rochester, NY

Over the years, top management broke down the functional departments of an organization into very specific operating units. This fostered an isolationism that spread to middle management and then to the floor-level supervisors. As every job became more narrowly defined and the "make it faster" philosophy continued, fingerpointing began between different departments.

Isolationism might have been acceptable when we couldn't make enough. Now we can't seem to make it well enough to satisfy the demands of a marketplace that prizes quality above price and delivery. While we were busy making it faster and sorting out the bad from the good, our international competitors were busy making it right the first time. If we are to improve quality and increase productivity, we need a way to bring our functional departments back together at the management level.

When the desire to make the product right the first time is paramount in the mind of everyone in an organization, the organization is naturally led to study the capability of the process which produces the product. Process capability studies lead immediately to the need for communication between the functional departments of the organization. Engineering, manufacturing and quality control departments must be in agreement on what the goals for the product are and must have an understanding of the difficulties encountered at each stage of

the process. Once agreement is reached on the methodology, SQC techniques can be used to determine the capability of the process.

Manual SQC methods

Most of the analytical tools useful for process capability studies are relatively simple to produce by hand and in fact, continue to be produced this way in Japan. But the Japanese have had many years to teach their work force how to produce and use them.

American top management has begun to rediscover these tools and their usefulness in making capital equipment, marketing, and design decisions. Discoveries such as this usually result in crash programs implemented at the shop-floor level without the required backup from the functional departments.

Quality circles came about due to the use of statistical techniques in Japanese manufacturing environments. Before these techniques reached the shop floor, top management, middle management, and floor-level supervisors had been exposed to the use of statistics as a management tool. Management was prepared to act on anything that a quality circle might have discovered to improve quality and productivity.

While the Japanese have been able to apply SQC techniques on a massive scale to their manufacturing environments using manual methods, we have not taken the time to train our work force in these methods. In order to apply SQC on the scale which will be required to make a significant impact on our quality and productivity problems, we must apply computer technology.

Process capability studies

A sequence of events must occur before you try to make process capability studies. The first thing you must do is determine if you really understand what it is you're trying to make. Does the design or drawing convey that to the people who have to make it? Is it possible to make the part the way it is drawn? Next, you must ask, can it be measured the way that the design people expect it to be measured? Then, are your measurement instruments equal to the task of measurement? This is all part of initially determining where and how you are going to apply statistical techniques.

One application of SQC is the determination of process capability through use of control charts. Once a process is running and you understand its capability, you can use these charts to monitor the process and spot things going wrong before they actually get so bad that the product being produced is unacceptable.

Control chart techniques are relatively simple to use and can be applied to just about any process. A key element is to decide where to use this method and where not to. The challenge is to design an experiment that will clearly reveal when assignable causes of variation have occurred.

Properly used, these techniques can be applied almost anywhere, although using them in extremely low volume processes may not be practical. A process with high scrap or rework costs presents a good opportunity to test the power of SQC.

At the other extreme, people who go overboard and try to apply SQC everywhere are probably not going to benefit as greatly as they would if they first predetermined where the major dollar benefits might be, and then applied SQC accordingly. These studies do take time and a little bit of work. We have seen people try to apply SQC indiscriminately without determining what the cost benefits might be. They would have been better off going slower and learning more about what SQC has to offer.

For example, a company has 20 diecasting machines. What they learn about one machine would probably be applicable to all of the machines. If they learn that closely controlling water cooling temperature, open-and-close dwell time, and material temperature produced quality parts on one machine, they might apply this knowledge to all of their machines.

This is not to say that studying one is a substitute for studying them all. Just that an analysis of one may be much

Demo lab at Hansford Data Systems shows key hardware ingredients in a top-quality SQC system: a coordinate measuring machine for accurate part measurements, a minicomputer (here, the Hewlett-Packard 1000/45) for rapid processing, a printer for tabulated outputs and a plotter for control charts.

more cost effective than indiscriminately analyzing them all.

Real time

The trends toward wider use of adaptive controls are part of the movement to reduce variability or stop the process before serious problems occur, like tool failure. The kinds of applications we have been asked to look at are systems where you can collect information from the floor, and based on knowledge gained through process-capability studies, flag problems before they occur.

Some people call this an early-warning system. When you know the capabilities of all the machines on your shop floor, not just a brand new adaptive one, you can establish as a long-term goal a system for spotting problems anywhere on your production floor. Setting up something like this requires a great deal of thought on how to get data into such a system, what kind of rules to establish and what kinds of outputs are required.

Some companies are starting to talk about a quality command center that would encompass real-time collection of quality information from the floor and display it on video screens with automatic flagging of problem areas. But we haven't really seen any plant with a fully functional system of this sophistication yet. Many people are talking about it and some have put in preliminary systems.

To input the necessary data to a com-

puterized SQC system, we will be seeing greater emphasis on instruments that can link directly into the system and many more video terminals on the shop floor or built into the machinery. Most state-of-the-art NC controllers have built-in video, and there may be ways to link information from the tool into a quality networking structure. Some of the CAD/CAM people are talking about putting terminals on the floor. Color graphics systems at the machine, showing setup operations, could similarly handle quality information.

Inputs

Unless we better control the environment in which the machines and people work, much of the data coming from the typical shop environment could be very misleading. For instance, there might be wild variability in the measurements based on the way temperatures change in manufacturing environments. It's impossible to assume that everything grows and shrinks at about the same rate. Machine tools of cast iron or steel may be machining aluminum or other materials. There are ways to compensate to some degree for this—the use of master parts, for example. In this case, the measurement system would look for correlation between the part data and the master and reject readings beyond reason. This technique is frequently used under controlled environmental conditions where measurement certainty is a requirement.

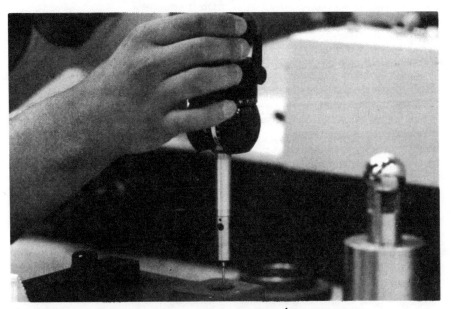

The coordinate measuring system, a Carl Zeiss WMM 850 CNC-CMM, is in an environmentally controlled room and provides the maximum in precision measurements of statistically selected parts from the process being monitored.

But to really automate factory data collection, we will have to better control the shop environment.

Short term, we still see a need for manual data input. But in the long run, as designers of equipment take into account the need to communicate with other computer systems, automatic data collection will result.

System requirements

Study your quality problems first, don't plan to completely automate quality data collection. You must decide what analysis is the most useful and what data collection is necessary to support it. The things you do very well may not need further measuring at all, while the things you do not do well may need 100 percent inspections.

Many of the companies we've talked to have never really thought about giving the quality people their own computer capabilities. In most cases the information needed to keep track of quality issues is much more detailed than that required by different elements of traditional management information systems. There was a study by the Aerospace Industries Association in 1978 in which they envision automated factories with a number of different computer complexes. One for engineering, one for manufacturing and one for quality. A quality information system could act as a filter to the rest of the organization, keeping in the esoteric 8-decimal point measurements while sending on to other depart-

ments the counts on acceptable parts.

A great deal of thought has been given to Computer Integrated Manufacturing (CIM) based on one enormous data base. Many experts say that's the way to go, but in reality there are a lot of data bases out there already, and people are not going to simply switch them over to a single giant data base. Very few computers are able to support all that in a single machine. The need for information should lead to greater acceptance for a distributed data base concept. This seems much more likely than one enormous "box" that you feed everything into. Also, you will need a real-time computer, rather than a batch machine, if you really intend to have an SQC early-warning system that responds to events as they happen.

Outputs

In one example, we chose 99 pieces from a die-cast machine run that produced 1500 pcs/day. Our mission was to pinpoint causes of some wild variability in the process. An operator kept a running list of everything that happened on the machine that particular day. When we ran the data from the sample parts through the computer, we could see the effects of each and every event that occurred during the run. For instance, the machine ran out of material and was refilled with material at a much higher temperature. This drastically changed how the process was operating. This change resulted in a shift for the rest of the production run that was clearly shown on the control chart.

As representative of this particular study, here are some of the key output plots produced by the computer system:
• **Probability plots.** This plot provides a quick visual check of the normality or validity of the data by plotting the samples' cumulative distribution function against a linearized normal probability scale. It shows the probability that the next measured point will be below a high-limit specification of 5.18000 mm. The closer the data tracks a best-fit

A good SQC program will force design and manufacturing people to communicate better: to preview prints, establish practical measurement criteria and make the design match the process capabilities.

straight line, the nearer it is to a normal distribution. It is also useful for estimating percentiles and percent out-of-specifications.

• **Histograms.** Most people are familiar with histograms. This plot can also be used to show if the data is normally distributed; however, it is a little more difficult to evaluate than the probability plot. It is commonly used to get a quick picture of the process once the process is in statistical control.

• **Control charts.** These answer the basic question of whether or not the process is in statistical control. In common use are X-bar and R or X-bar and S charts. R stands for range, the difference between the largest and smallest radial measurements in the selected subgroup of three consecutively produced parts. The range is very easy to calculate by hand. With a computer system, R may be replaced by S (Sigma), which is more meaningful statistically but more difficult to calculate. In this case, X-bar is the average radius of three consecutive parts in the same subgroup.

In this example, an interesting thing happened at Subgroup 18. The 0 signifies that the process went out of control (exceeded the upper control limit, Ucl). This occurred when the die-cast machine ran out of material and was refilled with material at a much higher temperature. Normally, we would expect the X-bar points to randomly vary about the central line. After the event at Subgroup 18, the variability between subgroups on the X-bar chart increased due to the change in material temperature. The Sigma chart remained essentially unaffected by the event and in control.

• **Normal inference chart.** This plot is a powerful chart used to make predictions about the process. Using information from the X-bar and S chart, we can estimate the parameters of the normal population and discover the natural tolerance of the process. For example, a customer wants to cut the print tolerance in half and marketing needs to know what to quote that customer for this increase in precision. With a chart like this, we can look at six standard deviations (± 3 sigma, 0.0125801 mm) and notice that that represents only 41.9 percent of the print tolerance (5.18000 − 5.15000 = 0.03 mm). So cutting that tolerance in half can be accomplished without changing the process. No capital equipment

decision would be necessary.

• **Polar deviation.** Mating parts may be well within size specs yet fail to go together because of location problems. The polar deviation plot shows observed center locations versus the specified center location. The object is to make all mating parts fall within the circle, the tolerance zone. This information can be used for tool adjustment decisions. This plot of centerpoints shows nearly everything in the fourth quadrant and illustrates the potential futility of first-piece inspection. Had this been a machining process and the single centerpoint in the first quadrant been from the first piece, we might have adjusted the machine in the minus y direction. If we did this, nearly all of the subsequent parts would have been bad. With all 99 centerpoints on the chart, we are able to make better decisions about how to adjust the tooling.

Do it yourself?

Many factors must be considered in the question of make or buy with regard to SQC software. The growing tendency seems to be that companies would rather buy off-the-shelf packages. One of the big reasons is that when software is internally generated, there is usually not time to pay enough attention to the human factors required to make the software user-friendly. Who uses the software, what kind of background they have, what terminology they are used to working with, how screen displays should look to them etc, are key issues that need to be addressed in designing user-friendly software. Most of the internally generated software we've seen is not very human oriented, or fails to provide the "hooks" to interact and react to necessary changes without rewriting the entire program. Or it fails to recognize

Probability plot

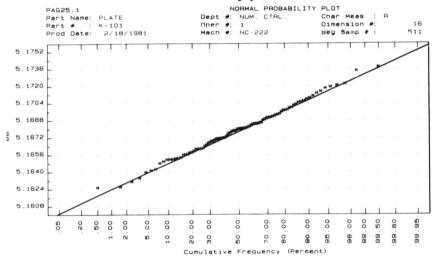

Histogram

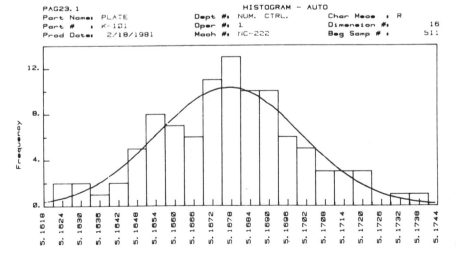

the need to communicate with systems already in place.

Another difficulty encountered with internally generated software is in the area of on-going support. Frequently, the people who wrote the software are not available to support it or the company has seriously underestimated the support required.

Companies will find themselves involved in all kinds of training when they start dealing with statistical quality control. The ability to quickly generate statistical information is only useful if the environment is there to act on the findings. The use of statistics can be very thought-provoking to an organization. Sometimes recommendations will evolve for changes in the way the company does business in certain areas. We see top management beating on desks, demanding quality, and putting in SQC training programs. But middle management is typically left out, thus minimizing the net value of this training to the company.

Management should use caution. They need to study their organization, use some of these statistical tools, and educate their people as they go along, rather than leap into massive programs. This is a long process to get effectively installed, whether by hand or with a computer system, and the process must be fully understood.

Implementation

Where are we as a country today in applying SQC techniques? Most of those who are using SQC are doing so manually. Those getting interested are saying "Let's educate the work force and have them do it," without realizing that they must do as much work with middle management to have them understand exactly how statistics can be used and what the benefits are.

One of the key advantages of a computerized SQC approach is the speed with which you can find out just where your company stands, where the problems are and where to start to work. With knowledge gained through process capability studies and related activities, you will find the lines of communications opening up between functional departments.

As functional departments begin to understand their responsibility for quality, the requirements will begin to emerge for an early-warning system approach to manufacturing. At this stage, distributed, real-time computer systems give us the tools we need to apply SQC effectively throughout the manufacturing environment. ∎

Control chart

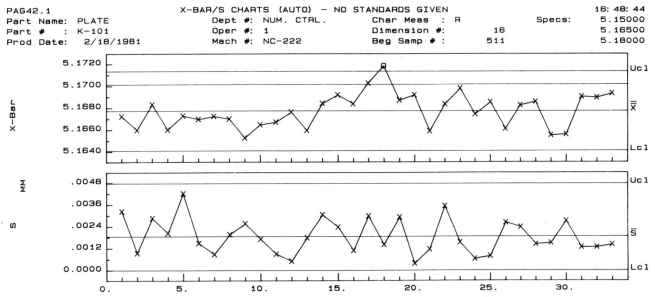

Normal inference chart

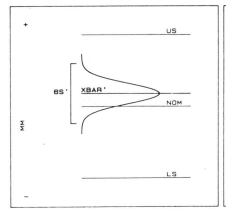

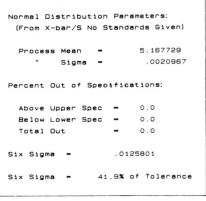

Polar deviation plot

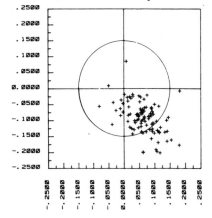

Presented at the SME AUTOFACT '87 Conference, November 1987

Engineering the 1990s Quality Function

Edward M. Stiles
Ed Stiles & Partners

Successful CIM implementation demands engineering the quality function into the automated production processes. Highly sophisticated facility engineering techniques must be used to incorporate inspection and quality assurance techniques into the computerized on-site controllers of processes, cells, and systems. On hundred percent inspection is the norm; the process will be right the first time; and first piece inspection will be eliminated. Data will be accumulated and analyzed, and corrective action taken automatically. Sample cases are used.

INTRODUCTION

One of the most important changes in the role of quality ontrol and assurance in the 1990's will be to truly engineer the quality function. Computer Integrated Manufacturing (CIM) demands engineering the quality functions into the automated production processes.

Recent developments in computer-controlled automated in-process inspection technology make it possible and profitable to install such equipment on even very high volume fastener production. Closed feed-back loops to CNC machines, or the use of automated statistical process control techniques at the work station greatly reduces, or even eliminates scrap. Operator measurements and calculations are not required. Rework is done away with because defective quantities are insignificant. Means of implementing these techniques are discussed in some detail. Two of many different installations are used to demonstrate the actions taken for success.

It is very realistic to expect the separate inspection function to disappear in the 1990's, along with the need to develop and apply statistical process controls and sampling plans to be used by an inspection function.

Quality professionals, teamed with manufacturing engineers, will apply highly sophisticated facility engineering techniques to insure the inspection and control techniques are built into the computerized on-site controllers of the production processes. 100% inspection will become an economic reality. The process truly will be right the first time. Even first-piece inspection will be done away with by automated set-up techniques.

Data will be accumulated, analyzed and real corrective actions executed without human interface. The whole body of manufacturing and quality engineering disciplines will change dramatically, to the point certification examinations and their supporting bodies of knowledge will be re-thought.

HISTORY OF INSPECTION ENGINEERING

Inspection Engineering developed over the past fifty years to attack three different areas of quality concern:

1. The high cost, or impossibility due to product destruction, of 100% acceptance inspection or testing. This led to the growth of sampling inspection, with a commitment to allow a percent of known defective product to pass through the system to the customer. On the other hand, it did reduce the high costs of inspection.

2. Attempts to control human error and judgement calls in accepting product, primarily through rigid inspection instructions and procedures.

3. Development of methods of capturing inspection data, for analysis and corrective action with, in the past few years, a growing emphasis on statistical process control.

This meant that a vast amount of engineering drawings, process instructions, specifications, past inspection data and a broad knowledge of information gathering techniques and measurement technology had to be combined into very detailed inspection instruction sheets or specifications to guide the activities of many inspectors. At the same time, it created a problem of preparing such instructions at a minimal cost by people with little formal technical or professional training. This problem was greatly compounded by the general belief of people assigned to such duties that they were in dead-end jobs, out of the mainstream of career paths to advancement in the quality profession or general manufacturing management.

Necessity forced the same solution that took place simultaneously in the manufacturing engineering areas of companies. Inspection Engineering narrowed its scope of effort in most companies to two activities. The first was the design and procurement of gaging and inspection equipment for use by inspection and quality control personnel. The second was the copying of old inspection instructions (as manufacturing engineers copied old process sheets) for use on newly introduced parts, or the revision thereof to accomodate engineering and process changes.

No time remained for new conceptual thinking, or for the intensive questioning of old inspection shibboleths. In particular, probably no inspection engineer ever considered the question:

WHY HAVE INSPECTION AT ALL?

If Manufacturing's true concern is to make all products right the first time, there is no need for a separate inspection function. Acceptance of each and every part must be an inherent part of the manufacturing process!

Consider the old arguments for sampling versus 100% inspection:

1. 100% inspection fatigues inspectors and leads to errors, accepting defective product.

2. Although sampling will accept lots with a certain percentage of defects, it is a forecastable item and can be tailored to meet the customers' requirements, if known.

3. 100% inspection as a separate process is far more expensive than sampling inspection.

4. The manufacturing process can be adequately controlled by sampling inspection and charting the data so obtained on control charts, applying the principles of statistical quality control (SPC).

Do these arguments meet the goal of 100% acceptable product made right the first time? Of course not!

THE FUTURE INPROCESS INSPECTION SYSTEM

In the 1990's, automation and computer control will alter basic manufacturing concepts drastically. As discussed later in this article, it is becoming possible to measure each unit of product during the manufacturing process and take corrective measures prior to producing the next piece. Carried to the ultimate, process "drift" due to tool wear, equipment problems and other assignable causes, or sudden process failures involving more than one part will become past history. In effect, this is 100% process control. The need for statistical process control is eliminated.

Instead of a process distribution, and historical data on which to base control charts, process history will be made up of a record of actual measurements made on every unit, plus the listing of corrections made, part by part. Human error will be eliminated by the utilization of self-correcting computerized numerical control (CNC) systems. The same CNC systems will track and report incipient equipment or tool failures for corrective action prior to actual break-downs and loss of product quality.

Inspection largely becomes an inherent portion of the basic manufacturing process. The costs associated with a separate inspection function will largely disappear. Inspection Engineering concerns will focus, in cooperation with the manufacturing engineers on the measurement capabilities, feedback loops and fool-proofing of the manufacturing processes.

At the same time, Manufacturing and Inspection Engineering become vitally concerned with the capturing of information of product measurements and process corrections. Process capability, based on computerized analysis of such data, will be used to establish new or improved processes, guide equipment selection and determine major maintenance and equipment overhaul/rebuilding programs. However, this process capability, unlike 1980's practice, will be based on a complete knowledge of every event that occurred in the process, rather than a hopefully random sample of events. Further, the data will be corrected to standard ambient conditions from actuals at the time it is captured.

Neither random sampling or true knowledge of event occurance is present in present SPC systems, for obvious reasons. Human operators take parts as they come to the machine, not in true sequence or at random. Human operators do not record all corections, re-setting of tools and other significant occurances in a process.

No adjustment is made of recorded measurements to reflect ambient conditions versus standard conditions.

Where available technology does not provide, apparently, for inprocess control, Manufacturing and Inspection Engineering will need a research and development function charged to develop and imbed such control in the process.

This will require professional, highly trained engineers in the Inspection Engineering function. At the same time, these people should become a much-sought-after technical resource in companies for quality assurance, manufacturing, and general management positions. This will solve much of today's problems with the perception of Inspection Engineering as a "dead-end" career, or a storage area for less competent people.

INSPECTION ALTERNATIVES IN THE 1990'S

To a certain extent, inspection is now an uneasy balance between inspection cost and the costs of defects being produced and shipped to customers. The falicy here is that the effect on customers of defects shipped usually is grossly understated. It applies to both manufacturing and service industries and functions. How can the cost weight be eliminated from the equation, or reduced by combining it with other functions?

Relative cost of a product or part, per se, has always been a major concern in controlling inspection and quality calculations, even if some Manufacturing Engineers disagree. Their concerns range from vendor quality control to product/service delivery, with detours into the material review board (MRB) along the way. Consider the following approaches, which reduce cost considerations to a large extent along the way:

1. An aerosol valve worth 0.7 cents.

   a. Automatically inspect 100% after each of five process operations in the automated assembly machime, and 400% test after last operation.

   b. Sort rejects by body mold identification as rejected.

   c. Accumulate data from (a) and (b) via the computer integrated manufacturing (CIM) system. Analyze for corrective action at the detail parts level in the Molding Department.

2. A critical aerospace fastener made//rom a $14,000 forging.

    a. At each operation, check part loading area for cleanliness with robotic vision and clean as required. If unable to do so, stop operation and call for a person.

Figure 1. Probe Verification of Part Location (1)

    b. Load part with robot, verify part position with CNC controlled probe, correct process controller instructions and tool positions for minor location errors, abort process for major errors.

    c. Probe each tool prior to cycle, correct process controller instructions and zero droop tool to correct position for mean dimensional cut, or replace tool with spare in magazine.

    d. Measure each characteristic generated, either with tool or probe and make requisite corrections.

    e. In the case of characteristics, such as relationships which cannot be checked in the machine, the robot unloading the part places it in a post-process automated gaging system, integrated with the CNC controller, which gages and corrects the CNC system while the next part is being loaded in the machine.

Figure 2. Robot Loading into Post-Process CNC Gage (1)

    f. Collect data on all corections, measurements, and abortions in the
        CNC controller for transmission through the CIM system for data
        reduction, analysis and part history logging. This is done on
        cell, factory control or the central mainframe computers.

Despite the wide disparity in product cost, does not each of the above
scenarios guarantee 100% outgoing quality at no, or very little inspection
cost, with little or no added process cost?

How many companies apply these techniques today in 1987?   There is much to
do in American industries to meet world-class quality levels by the 1990's.

1990'S MANUFACTURING AND INSPECTION ENGINEERS

In the 1990's, Manufacturing and Inspection Engineers will be:

1. Trained, and remain current in:
    a. CNC, robotics, and automation concepts
    b. CAD/CAM, and its uses in quality areas as well
    c. Research and development of new conformance control techniques
    d. Data communications, reduction and analysis
    e. Computing skills
    f. Process scheduling and control concepts

2. Conversant with both customer/user and suppliers quality systems and
    expectations.

3. Oriented to elimination as opposed to perpetuation of inspection as a
    separate function.

67

4. Cautious about introducing new techniques without full understanding of possible side effects. A new measuring advance is the sure way to stop production without improving outgoing quality.

5. Management and cost oriented, fully understanding the route to lowest cost is to "make it right the first time".

REQUIRED TECHNOLOGY FOR THE 1990'S

Areas of separate inspection activity that can be eliminated now with available proven technology include:
    Receiving Inspection
    Metal Removal
    Metal & Plastic Forming
    Material Status Alteration:
        Heat and other process treatments
        Coatings
        Surface deposits
    Joining:
        Adhesive
        Heat processes, including welding, brazing, etc.
        Adhesive and impregnation
    Assembly and Test
    Alternate Operations for the Above Basic Palanning

Areas where separate inspection activity cannot be elimated now include:
    Burring and Edge Treatments, Metallic or Non-metallic
    Dimensional Stack-ups of Multiple Parts
    Handling Damage
    Outside of Planned Processing Work
    Material Review Board Data Requests
    Inventory Purging
    Picking, Packing and Shipping

These are fertile areas for the Manufacturing and Inspection Engineering R & D functions to attack, as a joint team effort.

In this attack, past inspection history should serve as a guide. The constant question is "Why inspect this?". For example, a strong case can be made for checking one screw out of a shipment, if the screws are used in the in-house processes, to verify identity. Let the assembly machines reject the non-conforming screws with 100% automated inspection prior to insertion by the machines.

Likewise, one of the great breakthroughs in information processing in the 1970's was the concept of entering each piece of data into the systems only one time. The resultant data banks were then used in all programs and systems where they were required. The same concept should be applied in Inspection Engineering. Measure a piece characteristic once, and only once, then "remember" it forever.

MANAGEMENT DIRECTION FOR THE 1990'S

Plant, division, company and corporate management must take certain actions to create a suitable Inspection Engineering function in the 1990's:

Start now, with a planned program to tie quality needs into plans for automation and productivity improvement.

Support CIM, and integrated CAD/CAM operations.

Demand all facility purchases provide CNC operation, and communication capability under MAP/TOPS to support CIM operation.

Provide CAM access for quality and inspection engineers.

Recognize success requires evolution, not revolution; and that evolution starts today.

Recognize total cost in vendor selection, which is often far more for the lowest bidder, than selected higher bidders.

Provide internal audits to control vendor selection decisions.

Indoctrinate Purchasing in the Total Quality Concept.

Plan for and upgrade Inspection Engineers.

Consider eliminating Material Review, and any rework that is not to original specifications. Let the Production activity decide on scrap.

Eliminate Return-To-Vendor (RTV) decisions. You will only get the rejected material back, and not to specifications. Demand 100% good quality from all suppliers. In fact, consider automatic scrapping of defective supplier material whenever found in house.

Adopt a well-publicized policy of not paying for defective vendor material

EXAMPLES OF CURRENT TECHNOLOGY:

Some examples of current equipment and systems available for automated inprocess inspection include:

1. An automated inspection system that uses collimated light beams to inspect small parts, such as fasteners. It can be placed in line with automated hopper feeders on automated assembly machines. Rates up to 3,000 pieces per hour can be maintained. Tolerances can be held to 0.0004" or 0.0102 mm. The light beams can also be used to detect the absence of significant features, such as drive slots, sockets, and locking devices. It discards defects before they are fed to the assembly point, and keeps a defect history.

2. Suitable for transfer lines, CNC machines, or FMS systems, a statistical process management system can distinguish the source of complex information inputs from multiple machines; both from gaging devices and machine factors such as torque, pressure, temperature changes and horsepower. It automatically corrects for changes in ambient conditions. Results are timely correction of CNC or PC controller factors, tool compensation and replacement, etc. The system will display in real time traditional statistical process control data such as X-bar, R, R-bar, P, etc. In addition, it can also report on each operation's 200 previous inspection opertions in a Pareto analysis, either on a color display or a printer. These systems are currently in the process of being made MAP compatible.

3. Equipment is available, using a robot and lasers, to check that fasteners are present in an assembly, as well as other small parts, such as on PC boards. Again results are accumulated for analysis, either on line or through communication to a larger system.

4. Equipment is available, under CNC, to check either sheet or coil stock prior to feeding into the first operation of a metal-forming line. This equipment can be given vision capability to check the surface finish of the material, as for example, to meet aircraft quality requirements.

5. Variants of the equipment described in 4. above are being used in Western Europe to insure lanced or sheared parts to bereturned to the material for transport are arriving in fact at the next operation. Missing holes (as from broken punches) are also checked for quick problem detection.

6. Sensing systems are becoming available that measure the force with which one object is inserted into another. The advantages of such systems in the non-threaded fastener field are obvious, since the systems can stop the process and/or mark the offending part, keep data on occurances, and prepare the data for reports and analysis.

7. Electronic probes on machines can be programmed through the CNC's to measure O.D.'s, some I.D.'s, lengths, some relationships such as as concentricities, and proper part location prior to start of an operation.

8. Where on-machine probes cannot verify part characteristic relation-ships, post-process CNC gaging equipment can be used to determine actual relationships on the part just completed. The machine tool CNC controller can be adjusted by the post-process gage while the next part is being loaded into the machine by the robot.

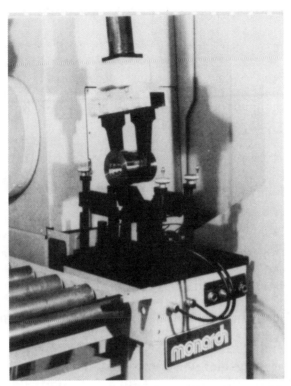

Figure 3. Post-Process CNC Gage for Concenticities   (1)

The characteristics common to all such examples are:
      Computerized control systems
      Automated feed and transfer of materials and parts
      Electronic, laser, or light operation for sensing
      Continued process flow
      100% inspection
      Removal from process or marking of discrepant materials
      Closed feedback loops
      Data collection, analysis, and reporting

From the foregoing, the need for Manufacturing and Inspection Engineers to be thoroughly conversant with computer systems, electronics, and advanced technology becomes obvious. Unfortunately, a very high percentage of today's Manufacturing and Inspection Engineers are from mechanical engineering fields. A considerable amount of new learning must be injected into these areas, either by training or hiring programs.

THE 1990'S MANUFACTURING QUALITY CHALLENGE

In most products, and in many service areas, few Americans will buy American products if Japanese or German products are available at an equivalent or even higher price. This is a concept many American executives still do not grasp. They remain wedded to the idea that Japan outsells competitive American products due to lower prices. This is no longer true.

Yet today certain American products enjoy fantastic quality reputations and are almost immune to inroads by overseas competitors. To name a few at random:
      Silverware
      Home appliances
      Photographic film and chemicals

Food & drink products, including wines

Obviously, many products "made in America" do not fall in this category, as yet. It can be done, and one of the simple ways to do it is by elimination of inspection through true ENGINEERING of the inspection processes which are then imbedded in the production processes.

Two interesting comments made recently illustrate all too well why we in American industry have not improved quality, including fastener quality, to the extent we could and should have over the past forty years, as both Japan and Western Europe did.

A Japanese engineer: "You Americans are too impatient. You demand everything right now. Solutions, product, ego satisfaction, all right now! You do not want to lose production time to correct a fault. You try to intuitively define what is wrong. You gather little data, and do little analysis. Without testing, usually you apply one solution. In the long run, you produce more bad product, and take longer, to solve and correct a fault...if you ever do succeed in fixing it."

A Danish professor of management: "Americans do not plan or experiment. By European standards, you do not specify accurately what your equipment is to do, and you almost cry when it doesn't do what you want. Then you sort the product, letting some bad parts go to your customers. Then you wonder why the world markets no longer respect American quality."

Those are strong criticsms of how the large majority of American manufacturers operate today. Change is required, one very importantchange being to reduce and ultimately eliminate separate inspection (sorting?) activities as outlined in this paper. "Make it right the first time 100% of time" must be the approach of American industry.

A true engineering approach to incorporating inspection into the process, rather than perpetuating it as a separate entity, is one of the basic steps to meeting this goal. With currently avaiable and the foreseeable technology there is no reason why American quality will lead the world at a competitive price. The major obstacle is the requisite change in management philosophies. The second obstacle is the education of engineering personnel to utilize the technology that is at, or soon will be, at hand. Both can be overcome, using the principles and approaches outlined herein.

_____

(1) All illustrations courtesy of The Monarch Machine Tool Co.

CHAPTER 2

STATISTICAL TECHNIQUES

Presented at the SME FABTECH WEST Conference, June 1984

Statistical Process Control: Fundamental Concepts

Jacob Frimenko, Ph.D.
Eastern Michigan University

Process Variation

In order to fully understand and utilize statistical process control (SPC) methods, it is necessary first to understand the concepts of process variation. Since every manufacturing operation contains many sources of variability, no two products, sub-assemblies, or readings of a process parameter taken from the same manufacturing process are <u>exactly</u> alike. The differences may be obvious or unmeasurably subtle, but they are always present. Differences may be observed for a specific measured characteristic (part specification - size dimension, material strength, material hardness, etc.) between two or more parts or between the readings of the manufacturing process parameters (process controls - cycle time, temperature, pressure, speed, feed, etc.) used to produce the parts. The values for individual measurements differ from eachother and are not predictable. Groups of measurements taken during a specific time period may be described as a distribution. The distribution of measured values is defined by an index of central tendency (usually the arithmetic mean) and an index of dispersion (usually the range or standard deviation). The indices of central tendency ($\bar{X}$) and dispersion (R or s) for a distribution are predictable overtime and allow estimations of future measurements taken from the same operation.

Product and process specifications are established as the limits for the production of acceptable products. The specification limits are concerned only with the total variation of the product or process measurements. Each measurement is compared to the specification limits and declared acceptable (within specifications) or not acceptable (outside the specification limits). The evaluations of the individual measurements do not consider the source or nature of the variation that contributed to the measured values. The variation contained in a manufacturing process however, may be divided into two components. The components are (1) common or natural variation, and (2) special (assignable) variation. Each source of process variation contains information that is helpful for managing the manufacturing process. Therefore, the first step required for controlling a manufacturing process is to segregate the sources or causes of the process variation. After the types of variation are separated, then more specific questions may be asked by the analyst as a method for resolving manufacturing problems.

Sources of Variation

Common variation may be defined as the difference between measured values that are inherent to the specific design and organization of the manufacturing process. The differences due to common causes are natural and expected for the process. In fact, common variation is planned and designed into the process by the decisions made by management. When the machinery is selected, the raw material characteristics are established, the level of personnel training determined, and the numerous other "management" decision settled upon; a level of difference between measured product and/or process characteristics is determined. These decisions are assimilated or designed into the process and serve as a constant source of deviations. Common variations in a process are stable, predictable, random in nature, and form a single distribution over time. When only common sources of variation are responsible for the differences in the measurements, the process is said to be in a state of statistical control. These random occurrences of variation are the responsibility of management and not the responsibility of the production department.

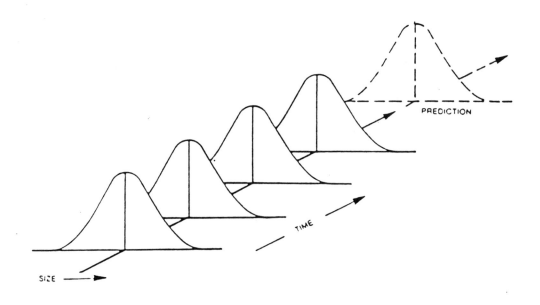

Figure 1. A process with only common variation is
stable over time and is predictable.
(Ford Motor Co., 1983, 4a)

Special causes of variation are traceable to specific factors and affect the manufacturing operation differently over time. The measurements of product characteristics or process parameters taken when special variation is present in the manufacturing system are not designed into the process and expected. Unless special causes of variation are identified and eliminated , they will affect the output of the process in unpredictable ways. A single distribution cannot adequately describe all of the measurements when special causes of variation are present. This unique variation is usually influenced and controlled by the decisions and actions of the production department personnel (i.e., operators, inspectors, foremen, superintendents, etc.). A manufacturing operation is called "out-of-control" when special sources of variation are present.

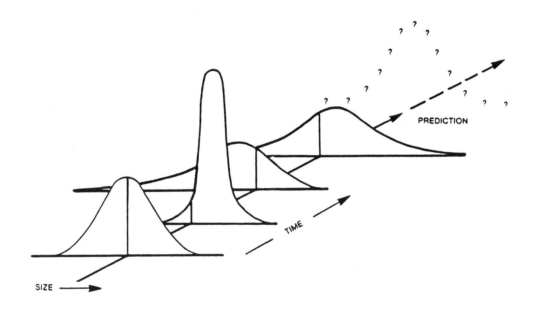

Figure 2. A process with common and special variation
is neither stable over time nor predictable.

(Ford Motor Co., 1983, 4a)

Tools and Techniques of SPC

The primary tool of a statistical process control program is the control chart. A control chart is a graphical aid that is used to analyze process variation. Although a large variety of control charts is available for the control of a production operation, the charts are generally classified according to the type of data they require. For the purposes

of SPC, data is classified into two categories. Variables data result
from measurements of a process or product characteristic using a continuous measurement scale. The data represents a level or the number of
units which describe the characteristics of interest (i.e., cycles per
second, degrees of temperature, size in millimeters). The units of a
variables measurement scale may be subdivided into fractions that are
meaningful. The control charts that employ variables data are concerned
with changes in both the central tendency (average) and the spread (range
or standard deviation) of the specific measurements being collected.
For this reason, two charts are required when variables data are employed.

Attribute data result from count-type or frequency measurements.
Only whole numbers (integers) are meaningful attribute data. The
collection of attribute data requires an evaluation or judgement that is
usually binary in nature (e.g., go/no-go, yes/no, acceptable/not acceptable). Control charts that require attribute data may be concerned
with controlling the number of defective parts being manufactured or with
the number of defects made per unit. Attribute control charts provide
a general or overall picture of product quality. Only one chart is
required for process control when attribute data are collected. The
charts most frequently used to control manufacturing processes are presented in Table 1.

Table 1. Commonly Used Control Charts and Their
Purposes

Data Type	Chart Name	Value Charted
Variables	$\bar{X}$ & R Chart	Sample means and ranges
	$\bar{X}$ & s Chart	Sample means and standard deviations
	$\bar{X}$ & Moving R Chart	Individual observations and moving ranges
	Median & R Chart	Sample medians and ranges
Attribute	p Chart	Proportion or percent of units defective per sample
	np Chart	Number of units defective per sample
	c Chart	Number of defects per inspection unit
	u Chart	Average number of defects per production unit

Statistical Control Conditions

A control limit is a line that is drawn on a control chart and is used as a critical value (99.73% confidence level) for evaluating the significance of process variation. The control limit identifies the range of values that could be expected if only common sources of variation were present in the process. This is the criterion for a process that is judged to be "in control."

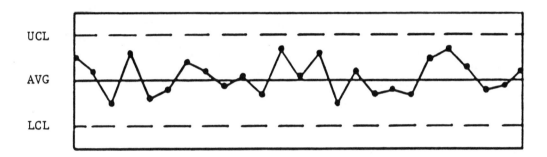

Figure 3. A process in control.

Points that are plotted beyond the control limits are considered extreme and indicate the presence of special variation. A single point beyond a control limit is sufficient for considering a process "out-of-control.

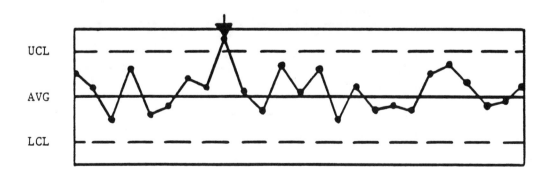

Figure 4. A process out-of-control - One sample point beyond a control limit.

When interpreting a control chart, the analyst must also consider the pattern of points as they are plotted within the control limits. The arrangement of points may define one or more condition that signal the presence of special variation. The plotted points may form four different patterns.

A run is defined as a series of data points (usually eight or more) that are constantly above or below the process center line. This pattern of points indicates a shift in the measured quality characteristic.

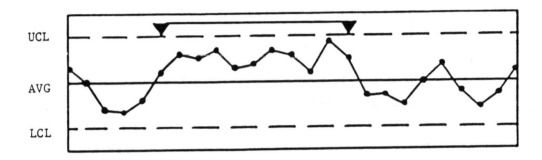

Figure 5.   A process out-of-control.- A run of data points on one side of the process average.

A trend is present when a series of data points (usually six or more) consistently increase or decrease. A trend reflects a gradual degradation or change in the measured quality characteristic.

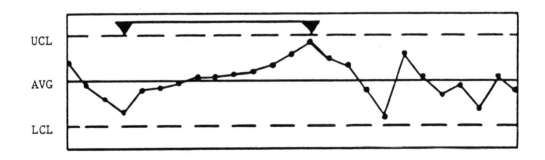

Figure 6.   A process out-of-control - A trend of sample points.

Cycles are repetitious patterns of the data points. These special causes of variation are often the more difficult causes of variation to identify and eliminate. Cycles result when psychological, chemical, mechanical, or seasonal factors affect the production operation.

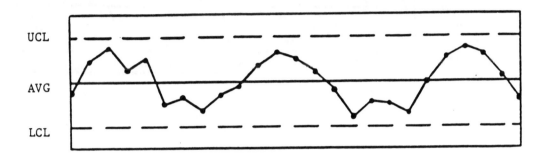

Figure 7. A process out-of-control - A cycle of sample points

Unusual patterns of variation may indicate a process stream effect or a fundamental change in the process. Unusual variation occurs when more tha 2/3 of the plotted points are in the center 1/3 of the chart or when more than 1/3 of the points are near the control limits.

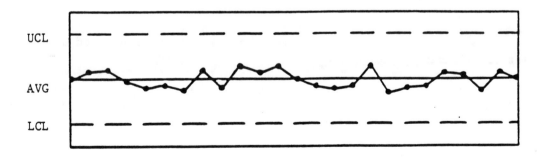

Figure 8. A process out-of-control - An excessive number of data points in the center third of the chart.

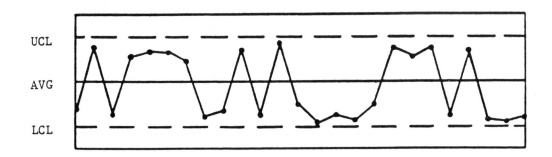

Figure 9. A process out-of-control - An excessive number
of data points close to the control limits.

Development of an $\bar{X}$ and R Chart

1. Determine the Sampling Frequency and the Sample Size.

   Before a process control chart may be constructed, data must be
collected. In order to accumulate data, three or more parts are usually
sampled from the manufacturing operation. The samples of pieces are
taken at logical intervals and inspected/measured for the characteristic
of concern. (Process parameters may also be sampled and used to control
the manufacturing operation. These measurements; speeds, feeds, cycle
times, temperatures, etc; often are more useful for controlling a process and for producing a quality product than are measurements taken
from the product itself.) The size of the sample is indicated by the
letter "n" and the number of samples with the letter "k." The sample size
(n) must remain constant for all samples for this charting method. The
sampling frequency should also be regular. The parts that comprise
the samples must come from only one process stream (i.e., a single tool,
head, die cavity, etc.).

2. Record Measurements on the Control Chart.

   Since both the $\bar{X}$ and R Charts must be used for variables data, both
of these chart forms and a data table are usually placed on the same
form. (Sample forms may be found in the Appendix.) The vertical scales
of the charts accomodate the $\bar{X}$ and R values and the horizontal scales
represent the sequence of samples across time, date, time, or shift from
which the sample was taken. The data values, the date or time, and the
plotted points should be aligned vertically.

3. Calculate the Mean ($\bar{X}$) and Range (R) for Each Sample.

The mean ($\bar{X}$) and range (R) must be calculated for each sample. These sample statistics will serve as indices of central tendency and dispersion for the process.

$$\bar{X} = \frac{X_1 + X_2 + \ldots + X_n}{n}$$

$$R = X_{high} - X_{low}$$

Where: X_1, X_2, ... are individual measurements in a sample
n is the sample size

4. Calculate the Process Center Line.

The average range ($\bar{R}$) and the process average (the average of the sample averages, $\bar{\bar{X}}$) serve as the process center line. These values are normally calculated only after thirty (30) or more samples/subgroups have been collected from the process. Since the range chart represents piece-to-piece variation, this chart is interpreted first.

$$\bar{R} = \frac{R_1 + R_2 + \ldots + R_k}{k}$$

$$\bar{\bar{X}} = \frac{\bar{X}_1 + \bar{X}_2 + \ldots + \bar{X}_k}{k}$$

Where : k is equal to the number of subgroups (or samples) and R_1, $\bar{X}_1$ are the respective range and average for the first sample, R_2, $\bar{X}_2$ the range and average for the second sample, etc.

5. Calculate Control Limits for the Charts.

Control limits indicate the amount of change sample averages and ranges would vary if only common causes of variation were operating in a process. The control limits are based upon the sample size and the amount of within sample variation reflected by the sample ranges. Constant values which vary by sample size (n) are used in the formulas for the control limits. The constants for samples ranging in size from two (2) to twenty-five(25) are presented in Table 2.

Table 2. Table of Constants and Formulas for Control Charts
(Ford Motor Co., 1983, 51)

Constants and Formulas

	$\overline{X}$ and R Charts*				$\overline{X}$ and s Charts*			
	Chart for Averages ($\overline{X}$)	Chart for Ranges (R)			Chart for Averages ($\overline{X}$)	Chart for Standard Deviations (s)		
Subgroup Size	Factors for Control Limits	Divisors for Estimate of Standard Deviation	Factors for Control Limits		Factors for Control Limits	Divisors for Estimate of Standard Deviation	Factors for Control Limits	
n	A_2	d_2	D_3	D_4	A_3	c_4	B_3	B_4
2	1.880	1.128	–	3.267	2.659	0.7979	–	3.267
3	1.023	1.693	–	2.574	1.954	0.8862	–	2.568
4	0.729	2.059	–	2.282	1.628	0.9213	–	2.266
5	0.577	2.326	–	2.114	1.427	0.9400	–	2.089
6	0.483	2.534	–	2.004	1.287	0.9515	0.030	1.970
7	0.419	2.704	0.076	1.924	1.182	0.9594	0.118	1.882
8	0.373	2.847	0.136	1.864	1.099	0.9650	0.185	1.815
9	0.337	2.970	0.184	1.816	1.032	0.9693	0.239	1.761
10	0.308	3.078	0.223	1.777	0.975	0.9727	0.284	1.716
11	0.285	3.173	0.256	1.744	0.927	0.9754	0.321	1.679
12	0.266	3.258	0.283	1.717	0.886	0.9776	0.354	1.646
13	0.249	3.336	0.307	1.693	0.850	0.9794	0.382	1.618
14	0.235	3.407	0.328	1.672	0.817	0.9810	0.406	1.594
15	0.223	3.472	0.347	1.653	0.789	0.9823	0.428	1.572
16	0.212	3.532	0.363	1.637	0.763	0.9835	0.448	1.552
17	0.203	3.588	0.378	1.622	0.739	0.9845	0.466	1.534
18	0.194	3.640	0.391	1.608	0.718	0.9854	0.482	1.518
19	0.187	3.689	0.403	1.597	0.698	0.9862	0.497	1.503
20	0.180	3.735	0.415	1.585	0.680	0.9869	0.510	1.490
21	0.173	3.778	0.425	1.575	0.663	0.9876	0.523	1.477
22	0.167	3.819	0.434	1.566	0.647	0.9882	0.534	1.466
23	0.162	3.858	0.443	1.557	0.633	0.9887	0.545	1.455
24	0.157	3.895	0.451	1.548	0.619	0.9892	0.555	1.445
25	0.153	3.931	0.459	1.541	0.606	0.9896	0.565	1.435

$$\text{UCL}_{\overline{X}},\ \text{LCL}_{\overline{X}} = \overline{\overline{X}} \pm A_2\overline{R}$$
$$\text{UCL}_R = D_4\overline{R}$$
$$\text{LCL}_R = D_3\overline{R}$$
$$\hat{\sigma} = \overline{R}/d_2$$

$$\text{UCL}_{\overline{X}},\ \text{LCL}_{\overline{X}} = \overline{\overline{X}} \pm A_3\overline{s}$$
$$\text{UCL}_s = B_4\overline{s}$$
$$\text{LCL}_s = B_3\overline{s}$$
$$\hat{\sigma} = \overline{s}/c_4$$

*From ASTM publication STP-15D, Manual on the Presentation of Data and Control Chart Analysis, 1976; pp 134-136. Copyright ASTM, 1916 Race Street, Philadelphia, Pennsylvania 19103. Reprinted, with permission.

$$UCL_R = D_4 * \bar{R}$$

$$LCL_R = D_3 * \bar{R}$$

$$UCL_{\bar{X}} = \bar{\bar{X}} + (A_2 * \bar{R})$$

$$LCL_{\bar{X}} = \bar{\bar{X}} - (A_2 * \bar{R})$$

Where: UCL_R is the Upper Control Limit for the range chart

LCL_R is the Lower Control Limit for the range chart

$UCL_{\bar{X}}$ is the Upper Control Limit for the averages chart

$LCL_{\bar{X}}$ is the Lower Control Limit for the averages chart

D_4, D_3, A_2 are constant values based upon the sample size

6. Determine the Vertical Scale for the $\bar{X}$ and R Charts.

The vertic scale for control charts should only be determined after the process center line and control limits have been calculated. If the scales are determined prior to these calculations, the process centerlines may be far from their expected position, the center of the chart. The control limits may also be located beyond the range of the charts' scale. The scale increments should be selected so the plotting of the sample points is relatively simple and easily understood (logical increments). The $\bar{R}$ value should be located about 1/3 to 1/2 of the distance up the vertical scale of the R chart. The UCL_R should be positioned approximately 2/3 to 3/4 of the distance up the scale. When the sample size is five (5) or less, no lower control limit is placed on the R chart.

The $\bar{\bar{X}}$ value should be approximately centered on the vertical scale of the $\bar{X}$ chart. The control limits should be centered about the $\bar{\bar{X}}$ value and should be spaced about 1/2 to 2/3 of the distance between the $\bar{\bar{X}}$ value and the end of the scale.

7. Plot the Sample $\bar{X}$ and R Values.

The $\bar{X}$ and R values for each sample must be plotted on their respective charts. The plotted points should be placed directly above the data and time sequence information contained in the data table. The plotted points may be connected with a solid line as an aid to the analysis of patterns and trends in the data. The connecting lines represent the time sequence for the sample data. The plotted points should be proofread and compared with the sample statistics before the control limits and process averages are drawn on the chart and the chart is analysed for process control.

8. Draw the Process Center Lines and Control Limits on the Charts.

   The $\bar{\bar{X}}$ and $\bar{R}$ lines should be drawn as   solid horizontal lines.
The control limits should be drawn as boldly dashed horizontal lines.
All lines should be labeled and dated. During the initial study phase,
the lines are considered trial center lines and control limits. These
may require revision as the process is modified and improved.

9. Analyze the Range Chart for Process Control.

   Since the control limits for both the range and averages charts are
based, in part, upon the $\bar{R}$ value, the R chart must be analyzed first.
If the piece-to-piece variation is not stable, the control limits
may not be appropriate and lead to false conclusions. The data points
must be analyzed by being compared to the control limits and by
being examined for unusual patterns ortrends. Remember, every process
must be evaluated against five (5) criteria before it may be judged
to be stable and "in control."

10. Analyze the Averages Chart for Process Control.

   The five (5) criteria for process control must be used to evaluate
the $\bar{X}$ chart. Identify and note all out-of-control conditions.

11. Identify and Eliminate Special Causes of Variation.

   The most critical   aspect of a SPC program is people's reaction
to the control chart signals. The identification and elimination of
special variation by production personnel requires an analysis of the
operation. The control chart itself often serves as a useful problem
solving tool. The timeliness of reaction, elimination, and prevention
of special causes of variation can never be over-emphasized. The reaction
time towards out-of-control conditions serves as the real criterion for
a production department's committment to understanding and controlling
their manufacturing processes and to the production of parts with a
consistent level of quality.

12. Extend the Control Limits for Ongoing Process Control.

   When the data points (at least 30) are consistently within the
control limits and in control, the control limits may be extended to
future samples of data. The production operator and the supervisor
should closely monitor the charts and provide prompt action to special
causes of variation. If the process is improved, or if extreme sources
of variation are eliminated, the process center lines and control limits

may no longer be appropriate. The changes made to the process may necessitate recalculation of the process center lines and the control limits.

Development of a p Chart

The p chart is used to monitor and control the percentage, or fraction, of defective units that are inspected. The inspected units may include a specified number of pieces (e.g., n=75) taken twice per shift or 100% of production grouped on an hourly, daily, or other logical frequency. Each part inspected must be evaluated and counted as being either acceptable or not acceptable (defective). Even if the part has more than one defect, only one tally may be added to the count of defective units.

1. Determine the Sample Size and Frequency.

The selection of sample size and frequency is very important when attribute data is used. shorter intervals allow faster feedback to the process if special causes of variation occur. Large sample sizes arealso desirable. Large sample siz es allow for more precise measurement of the process performance and are more sensitive to small changes in the process average. A sample of 100 units (n=100) could reflect changes in the process defective rate at increments of 1% (one defective unit out of 100 inspected units yeilds 1% defective, two defective units yields 2%). A sample of 50 units (n=50) could only reflect process changes that are greater than 2%.

Although it is desirable for the sample sizes to be identical from sample to sample, this need not be the case. Unequal sample sizes often result when the samples are based upon 100% of a shift's production. In the cases where the sample sizes vary, control limits should be calculated for the various different sample sizes. As an alternative to calculating control limits for each sample size, other advanced charting techniques are available.

2. Record Sample Data in the Data Table.

The number of defective units found in each sample and the sample size of each subgroup must be recorded in the data table of the attribute control chart form. Only one chart is required when attribute data is collected from the process.

3. Calculate the Proportion Defective for Each Sample.

The proportion defective for each sample must be calculated. This value is determined by using the following formula:

$$p = \frac{np}{n}$$

Where: p is the proportion defective for the sample

np is the number of defective units in the sample

n is the sample size

4. Calculate the Average Proportion Defective.

After thirty (30) or more samples have been collected, the average proportion defective may be calculated. The following formula is used:

$$\bar{p} = \frac{np_1 + np_2 + \ldots + np_k}{n_1 + n_2 + \ldots + n_k}$$

Where: $\bar{p}$ is the average proportion defective

np_1, np_2, ... are the number defective units in sample 1, sample 2, ...

n_1, n_2, ... are the sample sizes for subgroup 1, subgroup 2, ...

5. Calculate Control Limits.

The control limits for a p chart are calculated as the process average plus and minus an allowance that could be expected (99.73% confidence level) if only common variation was present for the sample. The size of the sample is a consideration for the calculations of control limits.

$$UCL_p = \bar{p} + 3\sqrt{\frac{\bar{p}(1 - \bar{p})}{n}}$$

$$LCL_p = \bar{p} - 3\sqrt{\frac{\bar{p}(1 - \bar{p})}{n}}$$

If the sample sizes are not constant and do not vary by more than 20%, the average sample size $(\bar{n})$ may be used as the typical sample size. When the $\bar{P}$ is very low or the sample size is very small, the lower control limit may be calculated as a negative number. In this instance, there is no lower control limit.

6. Determine the Vertical Scale.

The vertical scale of the p chart should indicate the proportion (or percent) defective and the horizontal scale the sequence of the samples. The $\bar{P}$ value should be located about 1/3 to 1/2 of the distance up the vertical scale. The upper control limit should be placed approximately 2/3 to 3/4 of the distance up the vertical scale.

7. Plot the Proportion Defective for Each Sample.

The p values for each sample must be plotted on the chart. The plotted points should be placed directly above the sample data. Lines may be drawn between the points as an aid for visualizing patterns and trends.

8. Draw the Process Center Line and Control Limits on the Chart.

The process average $(\bar{p})$ should be drawn as a solid horizontal line. The control limits should be drawn as dashed lines. If the sample sizes vary by more than 20% and if individual control limits were calculated for the different sample sizes, the control limits will be drawn as broken lines at different levels.

9. Analyze the Chart for Process Control.

The p chart must be evaluated against the five (5) criteria for process control.

      a. Points beyond control limits.
      b. Runs - 8 or more points on one side of the process average.
      c. Trends - 6 or more points increasing or decreasing.
      d. Cycles - repetitious patterns.
      e. Unusual variation.

10. Identify and Eliminate Special Causes of Variation.

The process average and control limits may require recalculation if the process is improved and modified.

Development of a c Chart for Defects

Defects are specific occurrences of a condition or characteristic which does not conform to specifications or other inspection standards. A single unit has the potential for more than one defect (e.g., a cast part may have three excessively large porosity pits, a cold shut, and a hot tear for a total of five (5) defects).

1. Determine the Sample Size and Collect Data.

The sample sizes for a c chart must be equal. This allows the differences in the plotted points to reflect changes in the level of quality rather than changes in exposure. The sample size should be sufficiently large to permit at least one defect per sample. The sample size and the number of defects uncovered in each sample should be recorded in the chart data table.

2. Calculate the Process Average.

The average number of defects per sample ($\bar{c}$) serves as the process average. The following formula is used for this purpose:

$$\bar{c} = \frac{c_1 + c_2 + \ldots + c_k}{k}$$

Where: $\bar{c}$ is the average number of defects per sample

$c_1, c_2, \ldots$ are the number of defects found in sample 1, sample 2,....

k is the number of sample or subgroups

3. Calculate the Control Limits.

The control limits for a c chart are claculated with the following simple formula:

$$UCL_c = \bar{c} + 3\sqrt{\bar{c}}$$

$$LCL_c = \bar{c} - 3\sqrt{\bar{c}}$$

4. Plot the Sample c Values, Draw the Process Average, and Draw the Control Limits.

The sample c points should be plotted directly above the sample data. Proofread the plots before the process average and control limits are drawn. The process average should be drawn as a solid horizontal line and the control limits as boldly dashed horizontal lines.

5. Interpret for Process Control.

Evaluate the chart against the five (5) criteria for process control:

   a. Points outside control limits.
   b. Runs – 8 or more points on one side of the process average.
   c. Trends – 6 or more points increasing or decreasing.
   d. Cycles – repetitious patterns.
   e. Unusual variation.

Process Capability

Once all the special causes of variation have been identified and eliminated, the capability of the process may be assessed. The fundamental question asked during a capability analysis is, "What percentage of the parts manufactured conform to the product specifications?" When variables data are collected, the answer to this question may be derived by conducting a supplementary analysis of the control chart data. The control chart alone cannot answer the question because the capability issue addresses the distribution of all parts being produced. The data points that are used on an $\bar{X}$ chart represent averages of parts that were taken from the process stream. Before the capability of the entire process can be evaluated, an estimate of the variation for the individual parts must be made. The distribution of the individual parts is always wider than the distribution of averages, which serves as the basis for the control chart. An estimate of variation for the individual parts $(\hat{\sigma}_x)$ may be made by using the average range $(\bar{R})$ of the samples and a constant value (d_2) which is based on the size of the samples.

$$\hat{\sigma}_x = \bar{R}/d_2$$

This estimate of variation is a standard deviation and may be used to calculate the proportion of the distribution that is outside the specification limits. The z-score formula is used to transform the specification limits into a value that indicates distance from the average of the observed measurements. By using the z-value for the specification limits as an entry in the density table for the normal distribution, the proportion of parts beyond the specification limits may be estimated.

p_z = the proportion of process output beyond a particular value of interest (such as a specification limit) that is z standard deviation units away from the process average (for a process that is in statistical control and is normally distributed). For example, if z = 2.17, p_z = .0150 or 1.5%. In any actual situation, this proportion is only approximate.

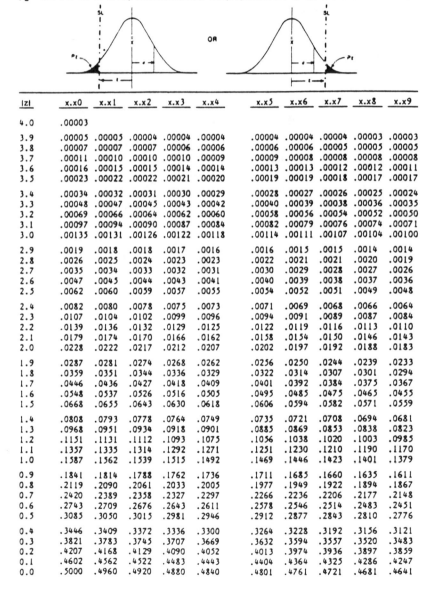

|z|	x.x0	x.x1	x.x2	x.x3	x.x4	x.x5	x.x6	x.x7	x.x8	x.x9
4.0	.00003									
3.9	.00005	.00005	.00004	.00004	.00004	.00004	.00004	.00004	.00003	.00003
3.8	.00007	.00007	.00007	.00006	.00006	.00006	.00006	.00005	.00005	.00005
3.7	.00011	.00010	.00010	.00010	.00009	.00009	.00008	.00008	.00008	.00008
3.6	.00016	.00015	.00015	.00014	.00014	.00013	.00013	.00012	.00012	.00011
3.5	.00023	.00022	.00022	.00021	.00020	.00019	.00019	.00018	.00017	.00017
3.4	.00034	.00032	.00031	.00030	.00029	.00028	.00027	.00026	.00025	.00024
3.3	.00048	.00047	.00045	.00043	.00042	.00040	.00039	.00038	.00036	.00035
3.2	.00069	.00066	.00064	.00062	.00060	.00058	.00056	.00054	.00052	.00050
3.1	.00097	.00094	.00090	.00087	.00084	.00082	.00079	.00076	.00074	.00071
3.0	.00135	.00131	.00126	.00122	.00118	.00114	.00111	.00107	.00104	.00100
2.9	.0019	.0018	.0018	.0017	.0016	.0016	.0015	.0015	.0014	.0014
2.8	.0026	.0025	.0024	.0023	.0023	.0022	.0021	.0021	.0020	.0019
2.7	.0035	.0034	.0033	.0032	.0031	.0030	.0029	.0028	.0027	.0026
2.6	.0047	.0045	.0044	.0043	.0041	.0040	.0039	.0038	.0037	.0036
2.5	.0062	.0060	.0059	.0057	.0055	.0054	.0052	.0051	.0049	.0048
2.4	.0082	.0080	.0078	.0075	.0073	.0071	.0069	.0068	.0066	.0064
2.3	.0107	.0104	.0102	.0099	.0096	.0094	.0091	.0089	.0087	.0084
2.2	.0139	.0136	.0132	.0129	.0125	.0122	.0119	.0116	.0113	.0110
2.1	.0179	.0174	.0170	.0166	.0162	.0158	.0154	.0150	.0146	.0143
2.0	.0228	.0222	.0217	.0212	.0207	.0202	.0197	.0192	.0188	.0183
1.9	.0287	.0281	.0274	.0268	.0262	.0256	.0250	.0244	.0239	.0233
1.8	.0359	.0351	.0344	.0336	.0329	.0322	.0314	.0307	.0301	.0294
1.7	.0446	.0436	.0427	.0418	.0409	.0401	.0392	.0384	.0375	.0367
1.6	.0548	.0537	.0526	.0516	.0505	.0495	.0485	.0475	.0465	.0455
1.5	.0668	.0655	.0643	.0630	.0618	.0606	.0594	.0582	.0571	.0559
1.4	.0808	.0793	.0778	.0764	.0749	.0735	.0721	.0708	.0694	.0681
1.3	.0968	.0951	.0934	.0918	.0901	.0885	.0869	.0853	.0838	.0823
1.2	.1151	.1131	.1112	.1093	.1075	.1056	.1038	.1020	.1003	.0985
1.1	.1357	.1335	.1314	.1292	.1271	.1251	.1230	.1210	.1190	.1170
1.0	.1587	.1562	.1539	.1515	.1492	.1469	.1446	.1423	.1401	.1379
0.9	.1841	.1814	.1788	.1762	.1736	.1711	.1685	.1660	.1635	.1611
0.8	.2119	.2090	.2061	.2033	.2005	.1977	.1949	.1922	.1894	.1867
0.7	.2420	.2389	.2358	.2327	.2297	.2266	.2236	.2206	.2177	.2148
0.6	.2743	.2709	.2676	.2643	.2611	.2578	.2546	.2514	.2483	.2451
0.5	.3085	.3050	.3015	.2981	.2946	.2912	.2877	.2843	.2810	.2776
0.4	.3446	.3409	.3372	.3336	.3300	.3264	.3228	.3192	.3156	.3121
0.3	.3821	.3783	.3745	.3707	.3669	.3632	.3594	.3557	.3520	.3483
0.2	.4207	.4168	.4129	.4090	.4052	.4013	.3974	.3936	.3897	.3859
0.1	.4602	.4562	.4522	.4483	.4443	.4404	.4364	.4325	.4286	.4247
0.0	.5000	.4960	.4920	.4880	.4840	.4801	.4761	.4721	.4681	.4641

When attribute data is used, the answer to the capability question is obtained relatively simply. For a p chart, the process capability is reflected by the average level of defective units ($\bar{p}$) when the process is in control. The percentage of units conforming to the product specifications may be derived by subtracting the $\bar{p}$ value from one (1) and multiplying by 100.

$$\text{Process Capability} = (1 - \bar{p}) * 100$$

The $\bar{c}$ value reflects the process capability for c charts and the number of defects per inspection unit. The process capability may be derived by subtracting the average number of defects per inspection unit ($\bar{c}$) from one (1) and multiplying by 100.

$$\text{Process Capability} = (1 - \bar{c}) * 100$$

Since the production department brought the process into a state of process control prior to the conduct of the capability study, only common or natural variation influences the findings. If the percentage of unacceptable parts is unsatisfactory, management must initiate action to address the problem. Since only common causes of variation are responsible for the excessive number of parts not conforming to specifications and only management have the authority and responsibility to determine the level of common variation, management must endeavor to improve the capability of the process.

BIBLIOGRAPHY

Burr, I. W. Statistical Quality Control Methods. New York: Marcel Dekker, 1976.

Deming, W. E. On some statistical aids towards economic production. Interfaces, 1975, 5, 1-15.

Ford Motor Company. Continuing Process Control and Process Capability Improvements. Statistical Methods Office, Operations Support Staffs, July, 1983.

Juran, J. M., Gryna, F.M. Jr., & Bingham, R.S. Jr. (Eds.). Quality Control Handbook. New York: McGraw-Hill, 1979.

Fundamentals of Analysis of Variance

CHARLES R. HICKS

Purdue University, West Lafayette, Ind.

Part I—The Analysis of Variance (ANOVA) Model

Ed. Note: This is the first in a series of three articles on the fundamentals of analysis of variance written by Prof. Hicks. Parts II and III will appear in subsequent issues.

Basically the analysis of variance technique is just what the name implies—partitioning the variance (i.e. the square of the standard deviation) of an experiment into parts in order to test whether or not certain factors introduced into the design of the experiment actually produce significantly different results in the variable tested. That is, for example, does the mold in which a casting was made affect the porosity of the casting? Does the scale used affect the weight of a vial of medicine? In each case, we are interested in testing whether the effect of the factor on the variable measured is significant when compared with the random variation in the process. Hence the F test or variance ratio is used to make such comparisons.

Let us consider a specific example adapted from some unpublished data collected at Purdue. Here the variable measured (say X) is the rate of fluid flow in cubic centimeters and we are interested only in the effect of one factor, nozzle type, on this rate of flow. The results

in Table I are for five runs through each of the three nozzle types.

The factor, nozzle type, is said to be in three categories as we have just three nozzles. It is assumed that these are the only nozzle types we are interested in. We do not wish to generalize our results to other nozzle types of which the three might be but a random sample. This

TABLE I—Number of c.c. of Fuel through three Nozzles for Five Trials.

		NOZZLE TYPE		
		A	B	C
Trial Number	1	96.6	96.6	97.0
	2	97.2	96.4	96.0
	3	96.4	97.0	95.0
	4	97.4	96.2	95.8
	5	97.8	96.8	97.0

is an important point: we are considering only these three nozzle types, so we have a factor in fixed categories. Had we been interested in these three nozzle types as a random sample of a whole population of nozzle types, nozzle types would be a random effect. In a one-way classification (one factor) like this one, the analysis used to get the results would be the same for either a random or a fixed effect, but the significance tests performed would be interpreted differently. This discussion will be confined to designs with fixed factors only.

For the one-way classification above, consider any observation—call it X_{ij}, where i is the number of the observation and j the category of the factor designation, i.e. the column in which X_{ij} is located. Thus $X_{23} = 96.0$ is the second observation for the third nozzle type. Now this number is affected by the category (or column) it is in, and also by some random error. We might then set up the model:

$$X_{ij} = \overline{X}' + B'_j + Z'_{ij}$$

where $\overline{X}'$ is a common term in all populations from which the experimental data were drawn, B'_j is a term designating the category or bias of the j factor from which the observation came ($j = 1, 2, 3$, in our problem) and Z'_{ij} is a random error. Our problem is to test the hypothesis that all three of these categories (or columns) came from like populations, or that: $B'_1 = B'_2 = B'_3 = 0$. If the least squares technique is applied to this model above, we find, for a sample of N observations:

$$X_{ij} = \overline{\overline{X}} + (\overline{X}_j - \overline{\overline{X}}) + (X_{ij} - \overline{X}_j)$$

where: $\overline{\overline{X}}$ is the best estimate of the population mean $\overline{X}'$.

$\overline{X}_j - \overline{\overline{X}}$ is the best estimate of the population B'_j.

and $X_{ij} - \overline{X}_j$ is the best estimate of the error or residual.

We note that this expression is a mathematical identity as both sides of the equation are equal for all values of X_{ij}.

Transposing in this identity:

$$(X_{ij} - \overline{\overline{X}}) = (\overline{X}_j - \overline{\overline{X}}) + (X_{ij} - \overline{X}_j)$$

which says that the deviation of each observation (X_{ij}) from the grand mean of all observations in the experiment $(\overline{\overline{X}})$ can be broken down into two parts: the deviation of the category mean from the grand mean plus the deviation of an observation from its own category mean.

Squaring both sides of this identity and summing first over all observations within a category and then over all k categories, the resulting expression is:

$$\sum_j \sum_i (X_{ij} - \overline{\overline{X}})^2 = \sum_j \sum_i (\overline{X}_j - \overline{\overline{X}})^2$$
$$+ 2 \sum_j \sum_i (\overline{X}_j - \overline{\overline{X}})(X_{ij} - \overline{X}_j) + \sum_j \sum_i (X_{ij} - \overline{X}_j)^2$$

or, $$\sum_j \sum_i (X_{ij} - \overline{\overline{X}})^2 = \sum_j \sum_i (\overline{X}_j - \overline{\overline{X}})^2$$
$$+ \sum_j \sum_i (X_{ij} - \overline{X}_j)^2$$

The middle term on the right hand side has vanished in the second expression since the summation of the cross product term equals zero. This expression says that the total sum of squares (of deviations) equals the sum of squares among category means plus the sum of squares within the categories. When each of these sums of squares (S.S.) is divided by the proper number of degrees of freedom (d. f.), the quotient represents an unbiased estimate of the population variance from which the data came. If our hypothesis is true that the nozzle types do not affect the mean rate of flow, then either the sums of squares among categories or the sums of squares within categories could be used to estimate the population variance, and each estimate is of the same variance, and the ratio of the two estimates will follow an F distribution. Hence an F test using the proper number of degrees of freedom will test the hypothesis that the means of all three categories are from the same population, or that $B'_j = 0$ for all j.

Now the actual calculations are not made by subtracting and getting the square of all these deviations but by making use of the binominal expansion of $\sum_i \sum_j (X_{ij} - \overline{\overline{X}})^2$. We can set up three rules for analyzing the effect of any factor on a given variable:

1) Square all the observations in the experiment and add, then subtract from this the square of the sum of all observations divided by N. This we call the total sum of squares. The sum of all observations squared and divided by N is often referred to as a correction term. Associated with the total S.S. are $N-1$ degrees of freedom.

Symbolically this rule is: $\sum_i \sum_j X^2_{ij} - T^2/N$,
where $T = $ grand total for the entire table.

2) Sum all observations for each category of a given factor, square this total and divide by the number of observations for this category, sum for all categories, then subtract the correction term as in (1). This is called the S.S. for the given factor and has $k-1$ degrees of freedom associated with it, where k is the number of categories given.

Symbolically this rule is: $\sum_j \left[\dfrac{(T_j)^2}{n_j} \right] - \dfrac{T^2}{N}$

where T_j is the total of all n_j observations in category j.

3) Subtract the S.S. for the factor (or factors) from the total S.S. This is the residual or error S.S. The degrees of freedom are $(N-1) - (k-1) = N-k$ for the design with just one factor.

The data are then summarized in a table. Let us try this on our rate of flow problem. It will simplify the calculations considerably if the data are first coded. This can be done by subtracting 96.0 from each observation and then multiplying each observation by 10. The results are all integers as shown in Table II.

Applying the rules, the total S.S. is 675.73. For the effect of the three nozzle types, applying step (2) we add all readings for each nozzle type, square, divide by five, and add for all three nozzles minus the correction term, giving 211.73. (Actually, here we divide by five after summing over all three types as the numbers of observations for each category are equal.) The remainder is $675.73 - 211.73 = 464.00$. This is the error

S.S. which is used as our "yardstick" to test the significance of the different nozzle types on the data, i.e. to test the hypothesis that $B'_1 = B'_2 = B'_3 = 0$.

The data are summarized at the bottom of Table II. The mean square is the unbiased estimate of the population variance found by dividing each S.S. by the associated number of degrees of freedom, and F is the ratio of the among nozzles mean square to the error mean square. In this problem F = 2.74. Consulting an F table with 2 and 12 d.f. the F necessary to claim a significant difference between nozzle types at the 5 percent level of significance is $F_{0.05} = 3.89$. Since our F is less than this, we conclude that our hypothesis cannot be rejected, and we behave as though the three nozzle types produce no differences in the mean rate of flow.

Now some bright boy in the organization notes that five different operators worked these three nozzles and the data could be analyzed further for possible differences in rate of flow due to the different operators. The problem now becomes an analysis of variance with a two-way classification of the data; i.e., two factors: nozzle type and operator, one in three categories (three nozzle types) and the other in five categories (five operators). Again we are assuming that the five operators are the only operators we are interested in, i.e. operators are a fixed factor. As each operator has worked with each nozzle type, we can analyze the data for differences in rate of flow among operators as well as among nozzle types. The model we assume now is:

$$X_{ij} = \overline{X}' + B'_j + C'_i + Z'_{ij}$$

where C'_i has been introduced to account for possible differences among operators where i = 1, 2, 3, 4, and 5. The analysis proceeds as before except for analyzing the five categories of the operator factor. In Table III, we have the analysis and the summary.

Here we applied the same rule (No. 2 above) to each operator—add for each operator, square, divide by three, and sum for all five operators minus the correction term.

$$\text{Symbolically:} \quad \sum_i \left[\frac{(T_i)^2}{k} \right] - \frac{T^2}{N}$$

This source of variation or S.S. among operators is also subtracted from the total S.S. leaving a different and "purer" error term than before. If we now recalculate the mean squares (that between nozzles stays the same), we can compare the among nozzles mean square with our revised estimate of the error mean square and also can compare the operator mean square with this error mean square. The F values are now as given in Table 3 along with their five percent significance F values. The results show that neither the nozzles nor the operators produce a significant difference in the average rates of flow even though we have now reduced the error term by accounting for another possible source of variation. In the first analysis (one-way classification), the operator effects were included in (that is, "confounded") with the error term. In actual practice, this other source of variation should have been foreseen in the original design and the second model used as shown in Table III. We thus fail to reject the hypothesis that: $B'_1 = B'_2 = B'_3 = 0$, and also fail to reject the hypothesis that $C'_1 = C'_2 = C'_3 = C'_4 = C'_5 = 0$.

TABLE III—Two-way Classification: Nozzle Types and Operators

		NOZZLE TYPE			
		A	B	C	Sums
Operator Number	1	6	6	10	22
	2	12	4	0	16
	3	4	10	−10	4
	4	14	2	−2	14
	5	18	8	10	36
	Sums	54	30	8	92

$$\text{Operator S.S.} = \frac{(22)^2 + (16)^2 + (4)^2 + (14)^2 + (36)^2}{3} - \frac{(92)^2}{15} = 185.06$$

Source of Variation	S.S.	d.f.	M.S.	F	$F_{.05}$
Total	675.73	14			
Among Nozzles	211.73	2	105.86	3.03	4.46
Among Operators	185.06	4	46.26	1.33	3.84
Error	278.94	8	34.87		

A further extension of our analysis would be possible if we were to repeat the same experiment, thus getting at least two observations for each operator-nozzle combination. This we call a replication and it is assumed that both (or all) replications are taken under the same conditions. Replication will enable us to analyze for a possible interaction between the two main factors, i.e. between nozzle types and operators. The replication would yield a still better estimate of the error in the experiment than was possible before. Of course this also requires more data. The model would now be:

$$X_{ij} = \overline{X}' + B'_j + C'_i + D'_{ij} + Z'_{ij}$$

where D'_{ij} represents the possible interaction between certain categories of one factor with categories of the other factor. Repeating the above experiment three times the results (coded) might be summarized as in Table IV.

TABLE IV: Nozzles vs. Operators Data with Replication

						Operator								
	1			2			3			4			5	
	Nozzle			Nozzle			Nozzle			Nozzle			Nozzle	
A	B	C	A	B	C	A	B	C	A	B	C	A	B	C
6	13	10	26	4	−35	11	17	11	21	−5	12	25	15	−4
6	6	10	12	4	0	4	10	−10	14	2	−2	18	8	10
−15	13	−11	5	11	−14	4	17	−17	7	−5	−16	25	1	24

TABLE II—Coded Data from Table I

		A	B	C	
Trial Number	1	6	6	10	
	2	12	4	0	
	3	4	10	−10	
	4	14	2	−2	
	5	18	8	10	
Sums:		54	30	8	92
Sums of Squares		716	220	304	1240

$$\text{Total S.S.} = 1240 - \frac{(92)^2}{15} = 675.73$$

$$\text{Nozzle S.S.} = \frac{(54)^2 + (30)^2 + (8)^2}{5} - \frac{(92)^2}{15} = 211.73$$

Source of Variation	S.S.	d.f.	M.S.	F	$F_{.05}$
Total	675.73	14			
Among Nozzles	211.73	2	105.86	2.74	3.89
Error	464.00	12	38.67		

Notice that this table has been set up somewhat differently than the usual 3 by 5 table as Table III. It is hoped that this will give a clearer idea of what interaction is and how its S.S. is computed. If we forget about nozzle types and operators for the moment and analyze the above as a one-way classification of 15 cells we find, applying the same rules set down earlier that:

$$\text{The total S.S.} = 7085.24$$
$$\text{Among cells S.S.} = 4047.24$$
$$\text{Within cells S.S.} = 3038.00$$

Now, the among-cells variation includes variation among operators and among nozzle types as well as chance variation. If we consider the S.S. for each of these main effects, we first calculate the operator S.S. by summing over the nine readings for each of the five operators (e.g. the total of the 1st three columns in Table IV gives this sum for operator 1 etc.) We find:

$$\text{Operator S.S.} = \frac{(38)^2 + (13)^2 + (47)^2 + (28)^2 + (122)^2}{9}$$
$$- \frac{(248)^2}{45} = 798.79$$

For nozzle types, we sum over the 15 readings for each of the three nozzle types (e.g. the sum for nozzle type A is the total of columns 1, 4, 7, 10 and 13 in Table IV) and find:

$$\text{Nozzle S.S.} = \frac{(169)^2 + (111)^2 + (-32)^2}{15} - \frac{(248)^2}{45}$$
$$= 1426.97$$

When these two main effect S.S. are subtracted from the among-cells S.S., we find that there is some variation left over (1821.48), which is unaccounted for by the nozzle types or by the operators. This is what we label the nozzle-operator interaction or N×O interaction S.S. This is a first order interaction as it is the simplest type observable in any experiment. It represents variation between the means of these "cells" not attributable to either of the main effects and is present because of the way certain operators might interact with certain nozzle types. For example, the second operator might run nozzle type A consistently too fast while running type C consistently too slowly and this would not show up as a nozzle or operator effect if another operator tended to reverse this by running nozzle C too fast and A too slowly. We now summarize for Table IV.

TABLE V: Summary of Nozzle, Operator Data with Replication

Source of Variation	Sum of Squares		d. f.	Mean Square	F	F.₀₅
Total		7085.24	44			
Among cells	Among Nozzles 1426.97 Among Operators 798.79 N × O Interaction 1821.48	4047.24	2 4 8	713.49 199.70 227.68	7.05 1.97 2.25	3.32 2.69 2.27
Within cells (error)		3038.00	30	101.27		

Referring to Table IV with its 15 cells, there are 14 degrees of freedom between these cell means. Now 2 d.f. may be accounted for by the three nozzle type differences, and 4 d.f. by the five operator differences which leaves: 14−2−4 = 8 d.f. for the interaction. In general, the interaction d.f. equals the product of the d.f. of the main effects producing this interaction. Here: (2)(4) = 8.

Whether to pool or not to pool the error and interaction for testing the main factors is still debatable among the experts, but we can test the significance of the interaction mean square with the error mean square as the "yardstick." If this is not significant, we can test the two main factors versus this error term or possibly pool the interaction and error terms as the yardstick for testing these main factors. The results above show the nozzle types producing a highly significant difference in mean rate of flow while operators show no significant differences, nor is the interaction quite significant if we use the five percent significance level.

In this discussion, we have tried to set up some general rules for getting the sums of squares and to review the analysis of variance technique. These general methods can easily be extended to analyze as many factors as you wish to assume in your original model, and also to compute the interaction of the first order or higher. It should be noted, however, that second order interactions are what is left in the among-cells S.S. (among cubes, really) after we subtract out the S.S. for the three main factors and all three first order interactions. The designs illustrated here are all factorial designs where each category of one factor is combined with each category of every other factor.

Finally, some mention should be made of the assumptions underlying the analysis of variance technique. The basis assumptions are:

1) The effect of all factors is additive. We have used linear models throughout this discussion.

2) The random errors were sampled from a normal universe.

Often some transformation of the variable can be made if this assumption is not met. How often we see a discrete variable used for analysis such as the number of defective items produced by several machines from several batches of raw material. Such data are likely to be from a binomial distribution, and the normality assumption is only approximately met if the average number of defectives is quite high. However, by using the percent defective as the variable and transforming it by an arc sine transformation, the normality assumption may be better satisfied.

3) The data must exhibit homoscedasticity or homogeneity of variance. That is, we should first show that no significant differences exist among the variances within the cells. This is usually tested with the Bartlett test. When transformations are made to induce normality, we may also expect more homogeneity of variance.

References

1) Griffith, Westman, and Lloyd, "Analysis of Variance" Parts I, II and III, *Industrial Quality Control*, Vol. IV, Nos. 5 and 6, March and May 1948

2) Bennett and Franklin, *Statistical Analysis in Chemistry and the Chemical Industry*, Chap. 7, Wiley, 1954

3) Ostle, Bernard, *Statistics in Research*, Ch. 9-12, Iowa State College Press, 1954

4) Anderson, R. L. and Bancroft, T. A., *Statistical Theory in Research*, McGraw-Hill, 1952

Part II—The Components of Variance Model and the Mixed Model

Ed. note. This is the second in a series of three articles on the fundamentals of analysis of variance. Part I appeared in the August issue, p. 17-20.

The setting up of a mathematical model and the determination of the Sums of Squares in the analysis of variance was emphasized in Part I. After the analysis, an F test was used to determine whether or not a particular factor in the model was present depending on how the variance estimate (M.S.) due to this factor compared in size with the estimate of error variance. Thus hypotheses on means were tested by a variance ratio or F test. In some experiments it is also of interest to estimate the components of variance or the amount of the total variance which might be attributable to the various factors in the model. However, one should only consider those factors which the F test has shown to have a significant effect on the measured variable.

In Part I, care was taken to discuss only fixed levels of a factor in the models. That is, the categories or levels of a given factor were considered the only ones of interest, whereas in practice some levels are often chosen at random from a large number of possible levels. When this is the case, the factor is said to be a random factor and a model based on such factors only is called a random model.

In industrial experimentation, some factors such as operators, days etc. might very well be considered in this way. The levels chosen, i.e. the particular operators, days etc. are but a random sample of a large number of levels that might have been chosen. Some statisticians feel that in these random models, one is chiefly interested in estimating the component of variance due to the given factor rather than in testing hypotheses on the equality of the mean values of the specific levels chosen. Thus, the random model has often been called the "Components of Variance" model and a model with all fixed levels of a factor the "Analysis of Variance" model[3].

Some authors even use different notation for the variance component for fixed levels and for random levels. In this discussion, the notation will be the same, but the distinction between the random and fixed levels will be made when the model is presented.

To get some idea of the importance of these components of variance, let us consider a problem in which the experimenter wished to study the effect of glass type and phosphor type on the brightness of a TV tube screen. The measured variable in this case was the current in microamperes necessary to produce a certain brightness: the larger this current the poorer the tube screen characteristics. The experimenter had only two glass types of interest and only three phosphor types. Thus, the factors were each at fixed levels and the model is similar to the problem considered in Part I. The model here is:

$$X_{ijk} = \overline{X}' + G'_i + P'_j + (GP)'_{ij} + \varepsilon'_{k(ij)}$$

where $\overline{X}'$ is the common effect

G'_i is the glass type effect and i = 1, 2 only.

P'_j is the phosphor type effect and j = 1, 2, 3 only.

$(GP)'_{ij}$ is the interaction which is possible between glass and phosphor types.

$\varepsilon'_{k(ij)}$ is the random error and k = 1, 2, 3 as it was decided to run three repeats of the experiment under the same conditions. The notation k(ij) simply means k different readings within, or nested in, each ij cell. The variance of these three gives an estimate of the variance of random errors.

The data and analysis of variance (ANOVA) are shown in Table I.

The analysis was actually run on coded data where 260 was subtracted from each observation and the result divided by five. Here both factors were considered fixed and a new column has been added to the ANOVA table called the "Expected Mean Square." Some refer to this as the 'components of variance' column and it is probably the most important part of the table. The symbols in this column are derived mathematically as the mean square to be expected if the model is correct. As their derivation is somewhat involved, it is our purpose to develop an easy method for a quick determination of these terms. For this problem, let us assume that the entries are correct, and see just what use can be made of the Expected Mean Square (EMS) column. There are two important uses to be made of this EMS column:

- To decide what F tests to run in the ANOVA.

- To estimate the components of variance due to each significant term in the model.

In Part I all F tests were performed by comparing the MS line of interest with the MS in the error line of the

TABLE I—Data and ANOVA for Tube Brightness Problem

Phosphor Type

Glass Type		A	B	C	
I		280 290 285	300 310 295	270 285 290	X_{ijk} = Current in microamps.
II		230 235 240	260 240 235	220 225 230	

Source	df	SS	MS	Expected Mean Square	
Total	17	618.44			
Between Glass	1	533.55	533.55	$\sigma_e^2 + 9\sigma_G^2$	(1)
Among Phosphors	2	49.77	24.88	$\sigma_e^2 + 6\sigma_P^2$	(2)
G × P Interaction	2	1.78	0.89	$\sigma_e^2 + 3\sigma_{GP}^2$	(3)
Error	12	33.31	2.78	σ_e^2	(4)

table. For a completely fixed model all effects are tested against the error MS or the error line [line (4) in Table I]. From the EMS column, it is clear that this should be so, as for example, to test the hypothesis that the glass type means $\overline{G}'_1$ and $\overline{G}'_2$ are equal is the same as testing whether or not the variance of these is zero or $\sigma_G^2 = 0$. If one assumes the hypothesis true; i.e., assumes $\sigma_G^2 = 0$, the ratio of line (1) to line (4) would be $\frac{\sigma_e^2}{\sigma_e^2}$ or 1. The departure from unity, being due only to chance, is checked by the F table for the degrees of freedom which apply. If, however, σ_G^2 is not zero, but the glass types really do effect the current, as they do here, this ratio $\frac{\sigma_e^2 + 9\,\sigma_G^2}{\sigma_e^2}$ should depart significantly from unity and it is reasonable to conclude that σ_G^2 is not zero so that glass type effect is real. In a similar manner, it is noted that all effects in this fixed model should be tested against line (4), the error line, to check the hypotheses that $\sigma_P^2 = 0$ and $\sigma_{GP}^2 = 0$. All this may seem very obvious to those who always test main effects and interactions against the error MS but actually this procedure is only valid for a fixed model and it will be shown that in a random model these are not the correct tests to run. It is thus the EMS column which indicates which tests should be run for testing hypotheses.

If now the proper F tests have been run and the significant factors determined, how does one estimate the components of variance? First, any variance component such as σ_{GP}^2 above which is not significant is eliminated (set = 0) in the EMS column. Then the observed MS values are best estimates of these terms in the EMS column. Setting the MS terms equal to the significant EMS terms, one may solve for the variance components of interest. In the example, glass type and phosphor type were both very significant but the interaction was not. Set up the equalities:

$$533.55 = \sigma_e^2 + 9\,\sigma_G^2$$
$$24.88 = \sigma_e^2 + 6\,\sigma_P^2$$
$$2.78 = \sigma_e^2$$

Thus, the best estimate of the error variance is 2.78 coded or $(25)(2.78) = 69.5$ in $(\mu a)^2$. Substituting in the 1st expression:

$$533.55 = 2.78 + 9\sigma_G^2$$
$$\text{or} \quad 9\,\sigma_G^2 = 530.77$$
$$\sigma_G^2 = 58.77 \quad [\text{or } 1474.3\,(\mu a)^2]$$

and substituting in the 2nd:

$$24.88 = 2.78 + 6\,\sigma_P^2$$
$$6\,\sigma_P^2 = 22.10$$
$$\sigma_P^2 = 3.68 \quad [\text{or } 92.0\,(\mu a)^2]$$

The estimate of total variance then due to all these sources is (in original units): $69.5 + 1474.3 + 92 = 1635.8$ of which $\left(\frac{1474.3}{1635.8}\right)$, or 90.1 percent, is attributable to glass types, $\left(\frac{92}{1635.8}\right)$, or 5.6 percent, to phosphor types and $\left(\frac{69.5}{1635.8}\right)$, or 4.3 percent, to error. Even though the phosphor type contributes only 5.6 percent of the variance, it cannot be ignored as it has already been

shown to be significant. It is obvious, however, that the glass type is the major source of variation.

Returning to the EMS column, it is seen that it provides the clue for what tests to run as well as how to estimate the components of variance due to the various sources in the model. If a model contains some random factors and some fixed factors, it is called a mixed model. Many industrial experiments are of this type, where one factor may be operators, and interest lies not in these particular operators but only in variation among operators of which the ones used are but a random sample. Now the decision as to whether a factor is at fixed or random levels is a practical one and not up to the statistician. If the experimenter considers operators as a random factor and wishes to generalize the results of his experiment beyond the few operators he chose, he must take measures to assure that the operators are really chosen at random from a large number of possible operators. If, in the example of Tables IV-V in Part I, operators are considered as randomly chosen and the nozzle types as fixed, the model is now a mixed model and the resulting analysis would be:

TABLE II—ANOVA for Nozzle, Operator Mixed Model Data

Source of Variation	df	SS	MS	EMS	
Total	44	7085.24			
Among Nozzles	2	1426.97	713.49	$\sigma_e^2 + 3\,\sigma_{NO}^2 + 15\,\sigma_N^2$	(1)
Among Operators	4	798.79	199.70	$\sigma_e^2 + 9\,\sigma_O^2$	(2)
N × O Interaction	8	1821.48	227.68	$\sigma_e^2 + 3\,\sigma_{NO}^2$	(3)
Error	30	3038.00	101.27	σ_e^2	(4)

From the EMS column it is seen that in order to test the hypothesis of no nozzle effect, or $\sigma_N^2 = 0$, the proper F test is to compare line (1) with line (3) instead of line (4)—the error line. A significant ratio here would indicate that the nozzle types did affect the rate of flow, whereas if line (1) is compared to the error line, it would not be clear whether the nozzle type or its interaction with operators, or both, produced the significant effect. The effect of operators and interaction is seen to be tested by the error line as in a fixed model. The decision then as to which tests to make after computing the MS column depends on what assumptions are made in the model as to which factors are at fixed levels and which at random levels.

The crux of the problem is then to derive these EMS values from the model for a given experiment. This would be a bit tedious if the mathematical derivation had to be carried out for every experiment. Some work on finite models by Cornfield[4] presented at the American Statistical Association meeting in Washington in 1953 and a more general formulation in a recent text by Bennett and Franklin[5] can be used to develop a set of rules for determining the EMS values. If one assumes that the levels of a factor, if random, are chosen from a large number (infinite) of possible levels these rules derived for a finite model can be applied here.

These rules are given below and are applied here to the foregoing problem:

1. Write the variable terms in the model as row headings in a two-way table.

2. Write the subscripts involved in the model as column headings. Indicate by F or R which subscripts correspond to fixed levels of a factor and which to random levels.

	F i	R j	R k	[Errors are always random].
Ex.				
N_i				
O_j				
$(NO)_{ij}$				
$\varepsilon_{k(ij)}$				

3. For each row (each term in the model), when a subscript letter is missing in the subscript of the row heading, copy in the table the number of observations under the corresponding column heading.

	i	j	k	
Ex.				
N_i		5	3	As: i = 1,2,3 nozzle types
O_j	3		3	j = 1,2,...,5 operators
$(NO)_{ij}$			3	k = 1,2,3 observations
$\varepsilon_{k(ij)}$				per cell.

4. For any subscript letter in the model which is in brackets [e.g. i and j in $\varepsilon_{k(ij)}$] place a 1 under the corresponding column heading. [e.g. under i and j here]:

	i	j	k
N_i		5	3
O_j	3		3
$(NO)_{ij}$			3
$\varepsilon_{k(ij)}$	1	1	

5. Fill in the remaining cells with a 0 or a 1 depending upon whether the subscript represents a fixed or random factor.

	F i	R j	R k
N_i	0	5	3
O_j	3	1	3
$(NO)_{ij}$	0	1	3
$\varepsilon_{k(ij)}$	1	1	1

6. To find the EMS for any term in the model:
Cover the entries in the column (or columns) which contain non-bracketed subscript letters in this term in the model. (e.g. for N_i: cover column i; for $\varepsilon_{k(ij)}$ cover column k etc.)

Multiply the remaining numbers in each row *if* their row heading (term in model) contains the subscript letter(s) of the term in question (e.g. for N_i: 5×3, 1×3, 1×1—not another 1×3, as O_j does not carry the subscript letter i in N_i).

These numbers (e.g. 15, 3, 1) now become coefficients of the corresponding variance terms in the EMS column.

	F i	R j	R k	EMS
N_i	0	5	3	$\sigma_e^2 + 3\sigma_{NO}^2 + 15\sigma_N^2$
O_j	3	1	3	$\sigma_e^2 + 9\sigma_O^2$
$(NO)_{ij}$	0	1	3	$\sigma_e^2 + 3\sigma_{NO}^2$
$\varepsilon_{k(ij)}$	1	1	1	σ_e^2

e.g. for the EMS of 1st line—cover column i as i appears in N. This leaves:

	i	j	k
N_i	0	5	3
O_j	3	1	3
$(NO)_{ij}$	0	1	3
$\varepsilon_{k(ij)}$	1	1	1

The product of the numbers left provide the coefficients for the corresponding variance components if we use only those containing i in the term of the model. Here these products are 15, 3, 1 (omitting the 3 at O_j as there is no i in O_j). These numbers are then the coefficients of σ_N^2, σ_{NO}^2, and σ_e^2 respectively for the nozzle type EMS line.

The proper tests to run are usually quite obvious from this EMS column. There is no necessity to assume anything about interaction before one tests any main effect because the F test follows whether the interaction term is significant or not.

Consider a second example in which an industrial engineering student wished to determine the effect of five different clearances of mating parts on the time required to position and assemble the part. As all such experiments involve operators, it is natural to consider operators as another important factor in the time required for executing these simple therbligs (elementary motion units). The measured variable here is the time required for position and assembly of the part and the experimenter also felt that whether the part was assembled directly in front of the operator or at arms length from the operator might influence this assembly time. He also wished to try four different angles in his set up from zero degrees directly in front of the operator thru 30, 60 and 90 degrees from this position. As all possible angles could be combined with both positions and all clearances and this arrangement tried by all operators, his was a factor, ' experiment where every level of each of four factors was combined with every other level of these factors. The experimenter also repeated the same set-up six times with every operator.

From this information, we see that there are four main effects: Operators (O_i) where he chose six operators at random, angles (A_j) chosen as fixed at 0, 30, 60 and 90 degrees, clearances (C_k) chosen at five fixed values, and locations (L_l) either in front of the operator or at arms length. Also there might be any or all pos-

TABLE III — EMS Determination on Time Study Problem

	df	R i	F j	F k	F l	R m	EMS
O_i	5	1	4	5	3	6	$\sigma_e^2 + 240\,\sigma_O^2$
A_j	3	6	0	5	2	6	$\sigma_e^2 + 60\,\sigma_{OA}^2 + 360\,\sigma_A^2$
$(OA)_{ij}$	15	1	0	5	2	6	$\sigma_e^2 + 60\,\sigma_{OA}^2$
C_k	4	6	4	0	2	6	$\sigma_e^2 + 48\,\sigma_{OC}^2 + 288\,\sigma_C^2$
$(OC)_{ik}$	20	1	4	0	2	6	$\sigma_e^2 + 48\,\sigma_{OC}^2$
$(AC)_{jk}$	12	6	0	0	2	6	$\sigma_e^2 + 12\,\sigma_{OAC}^2 + 72\,\sigma_{AC}^2$
$(OAC)_{ijk}$	60	1	0	0	2	6	$\sigma_e^2 + 12\,\sigma_{OAC}^2$
L_l	1	6	4	5	0	6	$\sigma_e^2 + 120\,\sigma_{OL}^2 + 720\,\sigma_L^2$
$(OL)_{il}$	5	1	4	5	0	6	$\sigma_e^2 + 120\,\sigma_{OL}^2$
$(AL)_{jl}$	3	6	0	5	0	6	$\sigma_e^2 + 30\,\sigma_{OAL}^2 + 180\,\sigma_{AL}^2$
$(OAL)_{ijl}$	15	1	0	5	0	6	$\sigma_e^2 + 30\,\sigma_{OAL}^2$
$(CL)_{kl}$	4	6	4	0	0	6	$\sigma_e^2 + 24\,\sigma_{OCL}^2 + 144\,\sigma_{CL}^2$
$(OCL)_{ikl}$	20	1	4	0	0	6	$\sigma_e^2 + 24\,\sigma_{OCL}^2$
$(ACL)_{jkl}$	12	6	0	0	0	6	$\sigma_e^2 + 6\,\sigma_{OACL}^2 + 36\,\sigma_{ACL}^2$
$(OACL)_{ijkl}$	60	1	0	0	0	6	$\sigma_e^2 + 6\,\sigma_{OACL}^2$
$\varepsilon_{m(ijkl)}$	1200	1	1	1	1	1	σ_e^2

sible interactions of these four factors. One can then write out the complete model as:

$$X_{ijklm} = \bar{X}' + O'_i + A'_j + (OA)'_{ij} + C'_k + (OC)'_{ik}$$
$$+ (AC)'_{jk} + (OAC)'_{ijk} + L'_l + (OL)'_{il}$$
$$+ (AL)'_{jl} + (OAL)'_{ijl} + (CL)'_{kl} + (OCL)'_{ik}$$
$$+ (ACL)'_{jkl} + (OACL)'_{ijkl} + \varepsilon'_{m(ijkl)}$$

where $i = 1, 2 \ldots 6 \quad k = 1, 2 \ldots 5 \quad m = 1, 2 \ldots 6$

$j = 1, 2 \ldots 4 \quad l = 1, 2$

Here operators were considered a random factor and all others were considered as fixed. Now we ask, "Just what tests can be made to check on the effect of angles, clearances, and locations on the assembly time and also how much of the variance might be due to operator differences which are almost certain to be found significant?" Also, "What interactions are important?"

Following the rules, we set down the model and determine the EMS column as in Table III.

From this EMS column it is easy to see what mean squares are to be compared to test any given hypotheses on either main effects or interactions. It is interesting to note that all fixed effects are tested by the term directly below them and all effects involving operators are tested by the error term.

These examples have shown the importance of the EMS line in an analysis of variance and how fixed and random models affect its determination.

Bibliography

(1) Griffith, B. A., Westman, A. E. R., and Lloyd, B. H., "Analysis of Variance," Parts I, II and III, *Industrial Quality Control*, Vol. IV, Nos. 5 and 6, March and May 1948

(2) Eisenhart, Churchill, "The Assumptions Underlying the Analysis of Variance", *Biometrics*, Vol. 3, No. 1, March 1947, p. 1-21

(3) Ostle, Bernard, *Statistics in Research*, Iowa State College Press, 1954, Ch. 9, p. 240-242

(4) Cornfield, Jerome, "Some Finite Sampling Concepts in Experimental Statistics," abstracted in *Journal of the American Statistical Association*, June 1954, p. 369-370

(5) Bennett, C. A., and Franklin, N. L., *Statistical Analysis in Chemistry and the Chemical Industry*, Wiley, 1954, Chap. 7

(6) Anderson, R. L. and Bancroft, T. A., *Statistical Theory in Research*, McGraw-Hill, 1952

Part III—Nested Designs in Analysis of Variance

Ed. note. This concludes Professor Hicks' article on the fundamentals of analysis of variance. Part I appeared in the August issue, p. 17-20, and Part II appeared in the September issue, p. 5-8.

Often in industrial experimentation an experimenter may confuse a factorial design, in which all levels of each factor are combined with all levels of every other factor, with another design known as a "nested" design. In this latter scheme, several levels of a factor may be nested within one level of another factor and be quite different for other levels of this same main effect. Recently an engineer asked about an experiment on a multi-headed machine in which there were four heads each supposedly doing the same job and he wished to study strain readings on each sample from each head and from five different machines. He argued that here was a case of five machines and four heads, so that a 5 × 4 factorial design and 20 experiments would provide the desired information if no repetitions were run within the 20 cells. If repeated four times, one has a 4 × 5 factorial design with four replications per cell, or 80 measurements. The model for this is:

$$X_{ijk} = \overline{X}' + M'_i + H'_j + (MH)_{ij} + \varepsilon'_{k(ij)}$$

where: $\overline{X}'$ is the common effect;

M'_i is the machine effect with i = 1,2,3,4,5;

H'_j is the head effect with j = 1,2,3,4;

$(MH)'_{ij}$ is the machine-head interaction effect;

$\varepsilon'_{k(ij)}$ is the random error with k = 1,2,3,4.

The data of Table I were recorded (after coding) and the analysis follows along the lines indicated in Part I where both heads and machines are considered fixed.

From this analysis one would conclude that there is no significant difference (at the 10 percent level) in the strain readings due to machines or due to heads but that there is a significant interaction between machines and heads.

Now a closer look at the problem indicates that each of the four heads was certainly not tried on all five machines, but actually the four heads were characteristic of the particular machine to which they belonged. Thus, there might be variation among heads but this would be within each machine as the four heads on another machine would, in fact, be entirely different heads. When such a situation occurs, the proper design is a nested design where the heads are nested within a machine, so the proper layout should be that shown in Table II.

In this table, the heads have been purposely numbered from 1 to 20 to emphasize that there really are 20 different heads to consider in this problem with four on each machine. This is essentially a one-way classification of data as there can be no interaction between ma-

TABLE I—Data and ANOVA for Sealing Machine Problem

	Machines					
		A	B	C	D	E
Heads	1	6 2 0 8	10 9 7 12	0 0 5 5	11 0 6 4	1 4 7 9
	2	13 3 9 8	2 1 1 10	10 11 6 7	5 10 8 3	6 7 0 3
	3	1 10 0 6	4 1 7 9	8 5 0 7	1 8 9 4	3 0 2 2
	4	7 4 7 9	0 3 4 1	7 2 5 4	0 8 6 5	3 7 4 0

Source	df	SS	MS	F	$F_{0.10}$*	EMS
Total	79	969.95				
Among Machines	4	45.08	11.27	1.05	2.04	$\sigma_e^2 + 16\,\sigma_m^2$
Among Heads	3	46.45	15.48	1.45	2.18	$\sigma_e^2 + 20\,\sigma_H^2$
M × H Interaction	12	236.42	19.70	1.84	1.66	$\sigma_e^2 + 4\,\sigma_{MH}^2$
Error	60	642.00	10.70			σ_e^2

(*10 percent level of significance used here)

TABLE II—Scaling Machine Problem—Nested Design

	Machine																			
	A				B				C				D				E			
	Head				Head				Head				Head				Head			
	1	2	3	4	5	6	7	8	9	10	11	12	13	14	15	16	17	18	19	20
	6	13	1	7	10	2	4	0	0	10	8	7	11	5	1	0	1	6	3	3
	2	3	10	4	9	1	1	3	0	11	5	2	0	10	8	8	4	7	0	7
	0	9	0	7	7	1	7	4	5	6	0	5	6	8	9	6	7	0	2	4
	8	8	6	9	12	10	9	1	5	7	7	4	4	3	4	5	9	3	2	0
Head Totals	16	33	17	27	38	14	21	8	10	34	20	18	21	26	22	19	21	16	7	14
Machine Totals	93				81				82				88				58			

chines and heads. The four heads are representative of a given machine and must be considered as heads from that machine only. The nested model can now be written as:

$$X_{ijk} = \overline{X}' + M_i' + H'_{j(i)} + \varepsilon'_{k(ij)}$$

with $i = 1, \ldots 5$; $j = 1, \ldots 4$; $k = 1, \ldots 4$.

The $\overline{X}'$, M_i' and $\varepsilon'_{k(ij)}$ are the same as in the first model and $H'_{j(i)}$ indicates true differences among heads (j's) within a given machine (i). The methods of Part II can now be used to determine the proper significance tests to be run on this design. If machines and heads are considered fixed as before, the resulting EMS column would be found as follows:

	F	F	R	
	i	j	k	EMS
M_i	0	4	4	$\sigma_e^2 + 16\,\sigma_m^2$
$H_{j(i)}$	1	0	4	$\sigma_e^2 + 4\,\sigma_{11}^2$
$\varepsilon_{k(ij)}$	1	1	1	σ_e^2

Here it is seen that one tests both the heads and machines vs. the error term. If, however, heads are considered random, then we would have:

	F	F	R	
	i	j	k	EMS
M_i	0	4	4	$\sigma_e^2 + 4\,\sigma_{11}^2 + 16\,\sigma_m^2$
$H_{j(i)}$	1	1	4	$\sigma_e^2 + 4\,\sigma_{11}^2 +$
$\varepsilon_{k(ij)}$	1	1	1	σ_e^2

Machines would be tested against the term for heads within machines. In either case, the results would be quite different from the original analysis as no interaction is possible in such a nested design.

In the computations of the sums of squares care must be taken in a nested design that the SS for heads within each machine be corrected by the correction term for that machine only and not by the overall correction term. Then these five terms may be combined to give the SS among heads within machines. For the example cited:

Total SS (as usual) $= 2990 - (402)^2/80 = 969.95$

Machine SS (as usual)
$$= \frac{(93)^2 + (81)^2 + (82)^2 + (88)^2 + (58)^2}{16} - \frac{(402)^2}{80}$$
$$= 45.08$$

Heads within Machines SS
$$= \frac{(16)^2 + (33)^2 + (17)^2 + (27)^2}{4} - \frac{(93)^2}{16}$$
(for first machine)
$$+ \frac{(38)^2 + (14)^2 + (21)^2 + (8)^2}{4} - \frac{(81)^2}{16}$$
(for second machine)
+ etc. for all 5 machines
$$= 282.87$$

Error SS (by subtraction) $= 969.95 - 45.08 - 282.87 = 642.00$, which is the same as in the previous analysis.

Thus, what was thought before to be head SS and machine-head interaction SS is really among-heads-within-machines SS, so the correct ANOVA is as shown in Table III.

The results of this correct analysis, when the design is recognized as a nested design, shows that at the 10 percent level of significance there is no significant machine effect, but there is a significant difference among heads within machines. Here then is evidence for more careful adjustment of the heads within the machine to obtain more homogeneous strain readings.

In some experimental work, the model considered may involve both a nesting within one or more factors, and other factors which are factorial (i.e. combined with all levels of the given factors). Such models would be referred to as "nested factorial" designs. The methods

TABLE III—ANOVA for Nested Scaling Machine Problem

Source	df	SS	MS	F	$F_{0.10}$	EMS
Total	79	969.95				
Among Machines	4	45.08	11.27	1.05	2.04	$\sigma_e^2 + 16\,\sigma_m^2$
Among Heads within Machines	15	282.87	18.85	1.76	1.60	$\sigma_e^2 + 4\,\sigma_H^2$
Error	60	642.00	10.70			σ_e^2

for determining the EMS column and hence the proper F tests to run are those cited in Part II as they allow for nested factors.

As an example of a nested factorial design, consider a problem by a former graduate student at Purdue who wished to improve on the number of rounds per minute on a Navy gun by a new method of loading the gun. His measured variable (X) was the number of rounds per minute and the main factor to be investigated was methods—the old vs. the new. However, all such loading depends on the team which loads the gun, so the investigator felt that the general physique of the men might have considerable effect on the resulting loading speed. Accordingly, he chose teams of men within three groups, where these groups were the slight men, average, and the heavy, more-rugged type. He was able to classify the men quite well in these three general groupings from Armed Services Classification tables. By a random choice of Navy ROTC students he was able to get three teams within each group which were thought to be representative of the group performance. It was felt that perhaps the slight-men (Group I) would do better on the new method than on the old method and that this might show up as a significant interaction between methods and groups. Considering all these factors, there were three things which might effect the number of rounds per minute: methods, groups, and teams within groups as well as some associated interactions and error. It was decided to make a repeat run on all of the teams to get a separate measure of error. The model can then be formulated as:

$$X_{ijke} = \bar{X}' + M'_i + G'_j + (MG)'_{ij} + T'_{k(j)} + (MT)'_{ik(j)} + \varepsilon'_{e(ijk)}$$

where

M'_i is the method effect, i = 1,2 (fixed effect);

G'_j is the group effect, j = 1,2,3 (fixed effect);

$(MG)_{ij}$ is the method-group interaction effect;

$T'_{k(j)}$ is the teams-within-group effect (a random effect), k = 1,2,3 for all j;

$(MT)'_{ik(j)}$ is the interaction effect between methods and teams;

$\varepsilon'_{e(ijk)}$ is the random error, e = 1,2.

The data and ANOVA are shown in Table IV.

The EMS values and proper F tests are found by applying the methods of Part II which leads to Table V.

To test the hypothesis of no method-team interaction; i.e. that $\sigma_{MT}^2 = 0$, the MT mean square is compared with the error mean square by the F test with 6 and 18 degrees of freedom or:

$F_{6,18} = 1.79/2.31$, which is less than one and hence not significant. The team effect is tested by com-

paring the team mean square with the error mean square:

$F_{6,18} = 6.54/2.31 = 2.83$, which is just significant at the five percent level when compared to 2.66 in the F table.

The methods-groups interaction may be tested by comparing the MG mean square to the MT mean square as indicated by the EMS entries in Table V. Here

$F_{2,6} = 0.605/1.79$, which is less than one and hence not significant. The group effect is tested vs. the teams-within-groups line or:

$F_{2,6} = 8.02/6.54 = 1.23$, which is below the F table value at the five percent significance level.

TABLE IV—Data and ANOVA for Gun-Loading Problem

	Group								
	I			II			III		
	Team			Team			Team		
	1	2	3	4	5	6	7	8	9
Proposed Method	20.2	26.2	23.8	22.0	22.6	22.9	23.1	22.9	21.8
	24.1	26.9	24.9	23.5	24.6	25.0	22.9	23.7	23.5
Present Method	14.2	18.0	12.5	14.1	14.0	13.7	14.1	12.2	12.7
	16.2	19.1	15.4	16.1	18.1	16.0	16.1	13.8	15.1

X = Number of rounds per minute

Source of Variation	df	SS	MS
Total	35	760.76	
Between Methods	1	651.95	651.95
Among Groups	2	16.05	8.02
M × G Interaction	2	1.19	0.60
Among Teams within Groups	6	39.26	6.54
M × T Interaction	6	10.72	1.79
Error	18	41.59	2.31

Finally the methods are tested vs. the MT interaction line, or:

$F_{1,6} = 651.95/1.79 = 364$, which is a highly significant value.

The results of the experiment then definitely show the superiority of the new method, which averages 23.58 rounds per minute as compared to 15.08 rounds per minute using the old method. There is also an indication of some variation in the average number of rounds per minute among the teams within the groups.

TABLE V—EMS Values on Gun-Loading Data

	F i	F j	R k	R e	EMS
M_i	0	3	3	2	$\sigma_e^2 + 2\,\sigma_{MT}^2 + 18\,\sigma_M^2$
G_j	2	0	3	2	$\sigma_e^2 + 4\,\sigma_T^2 + 12\,\sigma_G^2$
$(MG)_{ij}$	0	0	3	2	$\sigma_e^2 + 2\,\sigma_{MT}^2 + 6\,\sigma_{MG}^2$
$T_{k(j)}$	2	1	1	2	$\sigma_e^2 + 4\,\sigma_T^2$
$MT_{ik(j)}$	0	1	1	2	$\sigma_e^2 + 2\,\sigma_{MT}^2$
$\varepsilon_{e(ijk)}$	1	1	1	1	σ_e^2

This example shows an application of statistical methods involving both a nested and a factorial design in the same experiment. The methods for obtaining the EMS values, which in turn indicate the proper tests to run, are perfectly general and follow from the mathematical model assumed in the experiment. It is necessary to agree long before the experiment is actually run as to what the important factors are which may effect the measured variable and whether these factors are fixed or random and whether they are factorial with one another or may be nested levels of another factor. Once there is substantial agreement as to the nature of these factors, a mathematical model can usually be set up and the methods presented here can be used to indicate what analysis will follow from the model that has been chosen. It is assumed that randomization procedures are used in the actual ordering of the experimental conditions on the units tested.

Quality Levels in Acceptance Sampling

V. P. SINGH and H. R. PALANKI

IBM System Products Division, East Fishkill, Hopewell Junction, New York 12533

In attribute sampling, product quality levels may be determined by solving the well-known binomial equation. Tables giving the quality levels for specified probabilities of lot acceptance are generated for a series of sampling plans. The importance of these tables is demonstrated through some applications in industrial quality control. Attribute sampling plans similar to Military Standard (MIL-STD) and Dodge-Romig are derived, and confidence intervals for the population proportion defective are obtained.

Introduction

ATTRIBUTE sampling plans are used in many industrial applications of quality control to provide the assurance of specified quality levels. Two specifications are in general use, Acceptable Quality Level (AQL) and Lot Tolerance Percent Defective (LTPD). The AQL is usually defined as the maximum percent defective (or the maximum number of defects per hundred units) that for sampling inspection can be considered satisfactory as a process average; the LTPD is defined as a specified percentage defective representing the border line of distinction between satisfactory and unsatisfactory lots.

In this paper the acceptable quality levels are obtained by solving the binomial equation for known sampling plans. The quality levels for specified probabilities of lot acceptance are presented in tabular form for a series of given sampling plans. Several examples are then provided to demonstrate the use of these tables. It is shown that the attribute sampling plans based upon the binomial approximation can be easily obtained along the guidelines used in Military Standard (MIL-STD), Dodge-Romig, and other systems of sampling plans.

The advantage of this approach in obtaining the sampling plans is that it can be easily computerized, eliminating the need for sampling plan tables. Finally, the use of this procedure is shown for ob-

---

Dr. Singh and Mr. Palanki are advisory engineers in the IBM System Products Division at East Fishkill, New York.

KEY WORDS: Acceptance Sampling, Binomial Sampling Plans, Quality Levels.

taining confidence interval estimates of population proportion defective.

Theory

Assume that a manufacturing process is producing lots of quality p. Then, the distribution of the number of defectives, x, in samples of size n (with replacement) from this process follows the binomial law, viz.,

$$b(x, n, p) = \binom{n}{x} p^x (1 - p)^{n-x},$$
$$x = 0, 1, 2, \cdots, n \quad (1)$$
$$p\epsilon\,[0, 1]$$

It is to be pointed out that, in the case of sampling without replacement, (1) is an approximation to the hypergeometric law.

The probability, P_A, that a lot from this process will be accepted by a sampling plan is the probability that x does not exceed the acceptance number, c (the number of defectives allowed in the sample), which is given by (2),

$$P_A = \sum_{x=0}^{c} b(x, n, p). \quad (2)$$

From (2) it is obvious that P_A is a polynomial in p of, at most, degree n. Therefore, the evaluation of p for a given sampling plan (n, c) and for the probability of lot acceptance (P_A) involves finding the roots of an n-th degree polynomial. Since only those values of p are needed that lie in the interval zero to one, the problem reduces to finding an initial value of p in $[0, 1]$ closely satisfying (2). The required root of (2) will then be in the neighborhood of this p. To avoid guesswork, the following procedure was adopted to find an initial value of p.

Recall from (2),

$$P_A = \sum_{x=0}^{c} \binom{n}{x} p^x (1-p)^{n-x}$$

$$= 1 - \sum_{x=c+1}^{n} b(x, n, p)$$

$$= 1 - [1/B(c+1, n-c)] \qquad (3)$$

$$\int_0^p u^c (1-u)^{n-c-1} du,$$

where $B(a, b) = (a-1)!(b-1)!/(a+b-1)!$.

Let $\alpha = 1 - P_a$ and $I_p(a, b) = [1/B(a, b)]$

$$\int_0^p u^{a-1}(1-u)^{b-1} du. \qquad (4)$$

Then, from (3), we have

$$\alpha = I_p(c+1, n-c). \qquad (5)$$

Following Johnson and Katz [5], define y so that

$$y = t - \sum_{i=0}^{2} c_i t^i \Big/ \sum_{i=0}^{3} d_i t^i, \qquad (6)$$

where $t = [\log(1/\alpha^2)]^{1/2}$

$c_0 = 2.515517$, $c_1 = 0.802853$, $c_2 = 0.010328$,
$d_0 = 1$, $d_1 = 1.432788$, $d_2 = 0.189269$, $d_3 = 0.001308$.

Introduce

$$z = (y^2 - 3)/6$$

$$a = 1/(2c+1), \qquad b = 1/(2n-2c-1) \qquad (7)$$

$$h = 2/(a+b).$$

Then p may be approximated by the expression

$$p \simeq (1+c)/[1+c+(n-c)\exp(2w)], \qquad (8)$$

where

$$w = (y/h)(z+h)^{1/2} + (a-b)[z + (5/6) - (2/3h)]. \qquad (9)$$

Now, for given n, c, and α $(= 1 - P_A)$, the value of p (say, p^*) obtained from (8) is substituted in (2) to check whether this p^* is the required root. If p^* does not satisfy (2), several values from the neighborhood of p^* are tested in (2). If the difference be-

TABLE 1a. Quality Levels for $P_A = 0.025$

n/c	0	1	2	3	4	5	6	7	8	9	10
1	97.50	0.99									
2	84.19	98.75	0.99								
3	70.76	90.58	99.16	0.99							
4	60.24	80.60	93.24	99.38	0.99						
5	52.18	71.64	85.34	94.73	99.50	0.99					
6	45.93	64.13	77.73	88.20	95.68	99.59	0.99				
7	40.96	57.87	70.97	81.60	90.11	96.34	99.65	0.99			
8	36.94	52.65	65.09	75.52	84.30	91.48	96.82	99.69	0.99		
9	33.63	48.25	60.02	70.08	78.80	86.30	92.52	97.19	99.72	0.99	
10	30.85	44.50	55.61	65.25	73.77	81.29	87.85	93.33	97.48	99.75	0.99
11	28.49	41.28	51.78	60.98	69.22	76.63	83.26	89.08	93.99	97.72	99.77
12	26.46	38.48	48.42	57.19	65.13	72.33	78.91	84.84	90.08	94.51	97.92
13	24.71	36.03	45.45	53.82	61.44	68.42	74.87	80.78	86.15	90.92	94.96
14	23.16	33.88	42.82	50.81	58.12	64.88	71.14	76.97	82.34	87.24	91.62
15	21.80	31.95	40.47	48.09	55.11	61.63	67.73	73.42	78.74	83.67	88.18
16	20.59	30.24	38.36	45.65	52.39	58.67	64.58	70.13	75.36	80.25	84.81
17	19.51	28.69	36.45	43.44	49.91	55.97	61.68	67.09	72.19	77.03	81.56
18	18.53	27.30	34.71	41.42	47.65	53.49	59.02	64.27	69.26	73.98	78.47
19	17.65	26.03	33.14	39.59	45.58	51.21	56.56	61.65	66.51	71.14	75.56
20	16.84	24.88	31.70	37.90	43.68	49.11	54.29	59.23	63.95	68.49	72.81
21	16.11	23.83	30.38	36.35	41.92	47.18	52.18	56.97	61.57	65.99	70.23
22	15.44	22.85	29.16	34.91	40.29	45.38	50.23	54.88	59.35	63.65	67.80
23	14.82	21.96	28.04	33.59	38.78	43.72	48.41	52.92	57.27	61.46	65.51
24	14.25	21.12	27.00	32.36	37.39	42.16	46.72	51.10	55.33	59.41	63.36
25	13.72	20.36	26.04	31.22	36.09	40.72	45.14	49.39	53.50	57.48	61.34
26	13.23	19.65	25.14	30.16	34.87	39.36	43.66	47.79	51.79	55.67	59.43
27	12.77	18.97	24.30	29.16	33.74	38.10	42.27	46.29	50.19	53.97	57.64
28	12.34	18.36	23.51	28.23	32.67	36.91	40.96	44.88	48.67	52.36	55.94
29	11.94	17.77	22.77	27.36	31.67	35.78	39.74	43.55	47.24	50.84	54.33
30	11.57	17.23	22.08	26.53	30.73	34.73	38.58	42.29	45.89	49.40	52.82
31	11.22	16.71	21.43	25.76	29.84	33.73	37.48	41.10	44.62	48.04	51.38
32	10.89	16.23	20.81	25.03	29.00	32.80	36.45	39.98	43.41	46.75	50.01
33	10.58	15.76	20.23	24.34	28.21	31.91	35.47	38.92	42.27	45.53	48.72
34	10.28	15.34	19.68	23.68	27.46	31.06	34.54	37.91	41.18	44.37	47.48
35	10.00	14.92	19.16	23.06	26.74	30.27	33.66	36.95	40.14	43.26	46.31
36	9.74	14.53	18.66	22.47	26.07	29.51	32.82	36.03	39.16	42.21	45.19
37	9.49	14.17	18.20	21.91	25.42	28.78	32.03	35.17	38.22	41.21	44.12
38	9.25	13.81	17.76	21.38	24.81	28.10	31.26	34.34	37.33	40.25	43.11
39	9.03	13.49	17.33	20.87	24.23	27.44	30.53	33.55	36.47	39.33	42.13
40	8.81	13.16	16.93	20.39	23.67	26.81	29.84	32.79	35.66	38.46	41.20
41	8.60	12.86	16.54	19.93	23.14	26.21	29.18	32.07	34.88	37.62	40.31
42	8.41	12.57	16.17	19.48	22.63	25.64	28.54	31.38	34.13	36.82	39.46
43	8.22	12.29	15.81	19.06	22.14	25.09	27.94	30.71	33.41	36.05	38.64
44	8.04	12.03	15.48	18.66	21.68	24.57	27.36	30.08	32.72	35.31	37.85
45	7.87	11.78	15.15	18.27	21.23	24.06	26.80	29.47	32.06	34.60	37.09
46	7.71	11.53	14.84	17.90	20.80	23.58	26.26	28.88	31.43	33.92	36.37
47	7.55	11.30	14.55	17.55	20.39	23.11	25.75	28.31	30.82	33.27	35.67
48	7.40	11.07	14.26	17.21	19.99	22.67	25.25	27.77	30.23	32.64	35.00
49	7.25	10.86	13.99	16.88	19.61	22.24	24.77	27.24	29.67	32.03	34.35
50	7.11	10.65	13.72	16.56	19.24	21.82	24.32	26.74	29.12	31.45	33.73
60	5.96	8.94	11.54	13.93	16.20	18.39	20.51	22.57	24.59	26.58	28.53
70	5.13	7.71	9.95	12.02	13.99	15.90	17.73	19.53	21.28	23.02	24.72
80	4.51	6.78	8.75	10.58	12.31	13.99	15.62	17.20	18.76	20.29	21.80
90	4.02	6.04	7.80	9.44	10.99	12.50	13.95	15.17	16.77	18.15	19.49
100	3.62	5.45	7.04	8.52	9.93	11.29	12.61	13.89	15.17	16.41	17.63

tween the left-hand side and the right-hand side of (2) is positive (negative), the value of p^* is regarded as the highest (lowest) point of the neighborhood. Thus, only those values of p that are lower (higher) than p^* are substituted in (2), and the one satisfying the equation is taken as the desired root.

It is to be pointed out that, for $c = 0$, the value of p is found from

$$p = 1 - (P_A)^{1/n}. \tag{10}$$

Tables

From (2), (8), and (10), tables for the quality levels p are generated for the following values of n, c, and P_A:

$n = 1\ (1)\ 50\ (10)\ 100$

$c = 0\ (1)\ 10$

$P_A = 0.001,\ 0.0025\ (0.0025)\ 0.01,\ 0.02,\ 0.025$ $(0.025)\ 0.10,\ 0.10\ (0.05)\ 0.30,\ 0.30\ (0.10)$ $0.70,\ 0.70\ (0.05)\ 0.90,\ 0.90\ (0.025)\ 0.975,$ $0.98,\ 0.99\ (0.0025)\ 0.9975,\ 0.999.$

Each table corresponds to one of these specified values of P_A (the probability of lot acceptance) and all the values of n and c. The values of p (the quality levels) are expressed in per cents. Six examples of these tables are given, Tables 1a through 1f. (Others are available, upon request, from the authors.)

For a given sampling plan, $n = 20$, $c = 2$, and 95% probability of lot acceptance, the value of the quality level p is found as follows. In the left-hand (n) column of Table 1e the value 20 locates the designated row. In the top (c) row the value 2 locates the designated column. The entry at the intersection of the designated column and row gives the required value of p, 4.22%.

Applications

In this section we derive binomial sampling plans (BSPs) for various types of quality levels, such as AQL, LTPD, AOQL, and indifference quality (i.e., quality levels for $P_A = 0.5$). Attribute sampling plan tables keyed to these quality levels are available

TABLE 1b. Quality Levels for $P_A = 0.05$

n/c	0	1	2	3	4	5	6	7	8	9	10
1	95.00										
2	77.64	97.47									
3	63.16	86.47	98.30								
4	52.71	75.14	90.24	98.73							
5	45.07	65.74	81.07	92.36	98.98						
6	39.30	58.18	72.87	84.68	93.72	99.15					
7	34.82	52.07	65.87	77.47	87.12	94.66	99.36				
8	31.23	47.07	59.97	71.08	80.71	88.89	95.36	99.36			
9	28.31	42.91	54.96	65.51	74.86	83.13	90.23	95.90	99.43		
10	25.89	39.42	50.69	60.66	69.65	77.76	85.00	91.27	96.32	99.49	
11	23.84	36.44	47.01	56.44	65.02	72.88	80.04	86.49	92.12	96.67	99.53
12	22.09	33.87	43.81	52.73	60.91	68.48	75.47	81.90	87.71	92.81	96.95
13	20.58	31.63	41.01	49.46	57.26	64.52	71.30	77.60	83.43	88.73	93.40
14	19.26	29.67	38.54	46.57	54.00	60.96	67.50	73.64	79.39	84.73	89.60
15	18.10	27.94	36.34	43.98	51.08	57.74	64.04	70.00	75.63	80.91	85.83
16	17.07	26.40	34.38	41.66	48.44	54.83	60.90	66.66	72.14	77.33	82.22
17	16.16	25.01	32.62	39.56	46.05	52.19	58.03	63.60	68.92	73.99	78.83
18	15.33	23.77	31.03	37.67	43.89	49.78	55.40	60.78	65.94	70.88	75.60
19	14.59	22.64	29.58	35.94	41.91	47.58	53.00	58.19	63.19	67.99	72.61
20	13.91	21.61	28.26	34.37	40.10	45.56	50.78	55.80	60.64	65.31	69.80
21	13.29	20.67	27.06	32.92	38.44	43.70	48.74	53.59	58.28	62.81	67.19
22	12.73	19.81	25.95	31.59	36.91	41.98	46.85	51.55	56.09	60.48	64.75
23	12.21	19.02	24.92	30.36	35.49	40.39	45.10	49.64	54.05	58.32	62.46
24	11.73	18.29	23.98	29.23	34.18	38.91	43.47	47.87	52.14	56.29	60.32
25	11.29	17.61	23.10	28.17	32.96	37.54	41.95	46.22	50.36	54.39	58.32
26	10.88	16.98	22.29	27.19	31.82	36.26	40.54	44.68	48.70	52.62	56.43
27	10.50	16.40	21.53	26.27	30.76	35.06	39.21	43.23	47.14	50.95	54.66
28	10.15	15.85	20.82	25.42	29.77	33.94	37.97	41.87	45.67	49.38	53.00
29	9.81	15.34	20.16	24.61	28.84	32.89	36.80	40.60	44.29	47.90	51.43
30	9.50	14.86	19.53	23.86	27.96	31.90	35.70	39.39	42.99	46.51	49.94
31	9.21	14.41	18.95	23.15	27.14	30.96	34.67	38.26	41.77	45.19	48.54
32	8.94	13.98	18.39	22.48	26.36	30.08	33.69	37.19	40.61	43.94	47.21
33	8.68	13.59	17.87	21.85	25.63	29.25	32.76	36.18	39.51	42.76	45.96
34	8.43	13.21	17.38	21.25	24.93	28.46	31.89	35.22	38.47	41.65	44.76
35	8.20	12.85	16.92	20.69	24.27	27.72	31.06	34.30	37.48	40.58	43.63
36	7.98	12.51	16.47	20.15	23.65	27.01	30.27	33.44	36.54	39.57	42.55
37	7.78	12.19	16.05	19.64	23.05	26.34	29.52	32.62	35.64	38.61	41.52
38	7.58	11.89	15.66	19.16	22.49	25.70	28.80	31.83	34.79	37.69	40.54
39	7.39	11.60	15.28	18.70	21.95	25.09	28.12	31.08	33.96	36.82	39.60
40	7.22	11.32	14.92	18.26	21.44	24.50	27.47	30.37	33.20	35.98	38.71
41	7.05	11.06	14.57	17.84	20.95	23.95	26.85	29.69	32.46	35.18	37.85
42	6.88	10.80	14.24	17.44	20.48	23.42	26.26	29.04	31.75	34.42	37.03
43	6.73	10.56	13.93	17.06	20.04	22.91	25.69	28.41	31.07	33.68	36.25
44	6.58	10.33	13.63	16.69	19.61	22.42	25.15	27.81	30.42	32.98	35.49
45	6.44	10.11	13.34	16.34	19.20	21.95	24.63	27.24	29.80	32.31	34.77
46	6.30	9.90	13.06	16.00	18.80	21.51	24.13	26.69	29.20	31.66	34.08
47	6.18	9.70	12.80	15.68	18.43	21.08	23.65	26.16	28.62	31.04	33.41
48	6.05	9.51	12.54	15.37	18.06	20.66	23.19	25.65	28.07	30.44	32.77
49	5.93	9.32	12.30	15.07	17.71	20.27	22.74	25.16	27.54	29.86	32.15
50	5.82	9.14	12.06	14.78	17.38	19.88	22.32	24.69	27.02	29.31	31.56
60	4.87	7.66	10.12	12.42	14.61	16.73	18.79	20.80	22.77	24.72	26.63
70	4.19	6.60	8.72	10.71	12.60	14.43	16.22	17.96	19.68	21.36	23.03
80	3.68	5.79	7.66	9.41	11.08	12.69	14.27	15.81	17.32	18.81	20.28
90	3.27	5.16	6.83	8.39	9.88	11.33	12.73	14.11	15.47	16.80	18.12
100	2.95	4.66	6.16	7.57	8.92	10.23	11.50	12.75	13.97	15.18	16.37

TABLE 1c. Quality Levels for $P_A = 0.10$

n/c	0	1	2	3	4	5	6	7	8	9	10
1	90.00										
2	68.38	94.87									
3	53.58	80.42	96.55								
4	43.77	67.95	85.74	97.40							
5	36.90	58.39	75.34	88.78	97.91						
6	31.87	51.03	66.68	79.91	90.74	98.26					
7	28.03	45.26	59.62	72.14	83.04	92.12	98.51				
8	25.01	40.62	53.82	65.54	76.03	85.31	93.25	98.69			
9	22.57	36.84	49.01	59.94	69.90	78.96	87.05	93.92	98.84		
10	20.57	33.68	44.96	55.17	64.58	73.27	81.24	88.42	94.55	98.95	
11	18.89	31.02	41.52	51.08	59.95	68.23	75.95	83.08	89.52	95.05	99.05
12	17.46	28.75	38.55	47.53	55.90	63.77	71.18	78.13	84.58	90.43	95.48
13	16.23	26.78	35.98	44.43	52.34	59.82	66.91	73.63	79.95	85.84	91.20
14	15.17	25.07	33.72	41.70	49.20	56.31	63.09	69.54	75.68	81.49	86.91
15	14.23	23.56	31.73	39.28	46.40	53.17	59.65	65.85	71.78	77.44	82.80
16	13.40	22.22	29.96	37.12	43.89	50.35	56.54	62.50	68.22	73.71	78.96
17	12.67	21.02	28.37	35.19	41.64	47.81	53.74	59.45	64.96	70.27	75.39
18	12.01	19.95	26.94	33.44	39.60	45.50	51.18	56.67	61.98	67.12	72.08
19	11.41	18.98	25.65	31.86	37.75	43.40	48.86	54.13	59.25	64.21	69.02
20	10.87	18.10	24.48	30.42	36.07	41.49	46.73	51.80	56.73	61.52	66.18
21	10.38	17.29	23.40	29.10	34.52	39.73	44.77	49.66	54.42	59.05	63.56
22	9.94	16.56	22.42	27.89	33.10	38.12	42.97	47.68	52.28	56.75	61.12
23	9.53	15.88	21.52	26.78	31.80	36.63	41.31	45.86	50.29	54.62	58.85
24	9.15	15.26	20.69	25.75	30.59	35.25	39.76	44.16	48.45	52.64	56.74
25	8.80	14.69	19.91	24.80	29.47	33.97	38.33	42.58	46.73	50.80	54.77
26	8.48	14.15	19.20	23.92	28.42	32.77	37.00	41.11	45.13	49.07	52.93
27	8.17	13.66	18.53	23.09	27.45	31.66	35.75	39.74	43.64	47.46	51.20
28	7.89	13.19	17.91	22.32	26.55	30.62	34.59	38.45	42.24	45.94	49.58
29	7.63	12.76	17.33	21.60	25.70	29.65	33.49	37.25	40.92	44.52	48.06
30	7.39	12.36	16.78	20.93	24.90	28.74	32.47	36.11	39.68	43.19	46.63
31	7.16	11.98	16.27	20.30	24.15	27.88	31.50	35.05	38.52	41.93	45.28
32	6.94	11.62	15.79	19.70	23.44	27.07	30.59	34.04	37.42	40.74	44.00
33	6.74	11.28	15.33	19.14	22.78	26.30	29.73	33.09	36.38	39.61	42.79
34	6.55	10.97	14.90	18.60	22.15	25.58	28.92	32.19	35.40	38.55	41.65
35	6.37	10.66	14.50	18.10	21.55	24.90	28.15	31.34	34.46	37.54	40.56
36	6.20	10.38	14.12	17.62	20.99	24.25	27.42	30.53	33.58	36.58	39.53
37	6.03	10.11	13.75	17.17	20.45	23.63	26.73	29.76	32.74	35.67	38.55
38	5.88	9.85	13.41	16.74	19.94	23.05	26.07	29.03	31.94	34.80	37.62
39	5.73	9.61	13.08	16.33	19.46	22.49	25.44	28.34	31.18	33.97	36.73
40	5.59	9.38	12.76	15.94	19.00	21.96	24.85	27.67	30.45	33.18	35.88
41	5.46	9.16	12.46	15.57	18.56	21.45	24.27	27.04	29.76	32.43	35.07
42	5.33	8.95	12.18	15.22	18.14	20.97	23.73	26.43	29.09	31.71	34.29
43	5.21	8.75	11.91	14.88	17.74	20.51	23.21	25.86	28.46	31.02	33.55
44	5.10	8.55	11.65	14.56	17.35	20.06	22.71	25.30	27.85	30.36	32.84
45	4.99	8.37	11.40	14.25	16.98	19.64	22.23	24.77	27.27	29.73	32.16
46	4.88	8.19	11.16	13.95	16.63	19.23	21.77	24.26	26.71	29.12	31.50
47	4.78	8.03	10.93	13.66	16.29	18.84	21.33	23.77	26.17	28.54	30.87
48	4.68	7.86	10.71	13.39	15.97	18.47	20.91	23.30	25.66	27.98	30.27
49	4.59	7.71	10.50	13.13	15.66	18.11	20.50	22.85	25.16	27.44	29.69
50	4.50	7.56	10.30	12.88	15.35	17.76	20.11	22.42	24.69	26.92	29.13
60	3.76	6.33	8.63	10.80	12.88	14.91	16.89	18.84	20.75	22.64	24.51
70	3.24	5.44	7.42	9.30	11.10	12.95	14.56	16.24	17.90	19.54	21.15
80	2.84	4.78	6.52	8.16	9.74	11.28	12.79	14.28	15.74	17.18	18.60
90	2.53	4.25	5.81	7.27	8.69	10.06	11.41	12.73	14.04	15.33	16.60
100	2.28	3.83	5.23	6.56	7.83	9.08	10.29	11.49	12.67	13.84	14.99

in the literature, viz., MIL-STD-105D plans [6] for AQL, Dodge-Romig plans [4] for AOQL and LTPD, and Philips plans [3] for the indifference quality. These published tables use either a Poisson approximation or an arbitrary method of obtaining the sampling plan parameters. The approach used here differs in three respects from that supporting existing sampling plans: first, it uses only binomial approximation; second, it avoids using ranges of quality levels and lot sizes, thereby giving sampling plans for any specified quality level and for any given lot size; and, third, it lends itself to easy computerization. It is hoped that the third aspect of this approach will eliminate the need for sampling plan tables.

BSP for Specified AQL Values

For specified AQL values MIL-STD-105D sampling plans were developed with producer's risk (α) in the range 0.01 to 0.10 and sample size (n) related to a range of lot sizes. For normal inspection we have obtained Figure 1 by using the following relationship between the sample size (n) and the lot size (L):

$$n = \exp\left[\sum_{i=0}^{4} k_i (\log L)^i\right],$$

where $k_0 = -0.4826$, $k_1 = 0.2940$, $k_2 = 0.16404$, $k_3 = -0.0204$, and $k_4 = 0.000717$. The following is a step-by-step procedure to obtain BSP for a given lot size and AQL specification.

1. For a given producer's risk α, obtain a set of sampling plans from the table (of Tables 1a through 1f) with the heading Quality Levels for $P_A = 1 - \alpha$.
2. Obtain from Figure 1 the sample size (n) corresponding to the appropriate lot size (L).
3. For the sample size (n) obtained in step 2, find the closest sampling plan in step 1.

Example

Suppose AQL = 4%, $L = 300$, and $\alpha = 0.05$.
1. The set of sampling plans from Table 1c is shown in Table 2.

TABLE 1d. Quality Levels for $P_A = 0.90$

n/c	0	1	2	3	4	5	6	7	8	9	10
1	10.00										
2	5.13	31.62									
3	3.45	19.58	46.41								
4	2.60	14.26	32.05	56.23							
5	2.09	11.22	24.66	41.61	63.09						
6	1.74	9.26	20.09	33.32	48.96	68.12					
7	1.49	7.88	16.96	27.86	40.38	54.74	71.96				
8	1.31	6.86	14.68	23.96	34.46	46.18	59.37	74.99			
9	1.16	6.08	12.95	21.04	30.09	40.06	50.99	63.16	77.42		
10	1.05	5.45	11.58	18.76	26.73	35.42	44.83	55.04	66.21	79.43	
11	0.95	4.94	10.48	16.92	24.05	31.77	40.05	48.92	58.48	69.97	81.11
12	0.87	4.52	9.57	15.42	21.87	28.82	36.22	44.10	52.47	61.45	71.25
13	0.81	4.17	8.80	14.16	20.05	26.37	33.09	40.17	47.65	55.57	64.02
14	0.75	3.87	8.15	13.09	18.51	24.32	30.45	36.91	43.69	50.80	58.30
15	0.70	3.60	7.59	12.19	17.20	22.56	28.22	34.15	40.35	46.83	53.60
16	0.66	3.37	7.10	11.38	16.06	21.04	26.29	31.78	37.50	43.45	49.65
17	0.62	3.17	6.67	10.68	15.06	19.72	24.61	29.72	35.03	40.55	46.26
18	0.58	2.99	6.29	10.06	14.18	18.55	23.14	27.92	32.88	38.02	43.32
19	0.55	2.83	5.95	9.51	13.39	17.51	21.83	26.33	30.98	35.79	40.75
20	0.53	2.69	5.64	9.02	12.69	16.59	20.66	24.91	29.29	33.81	38.47
21	0.50	2.56	5.37	8.58	12.06	15.75	19.62	23.63	27.78	32.05	36.44
22	0.48	2.44	5.12	8.17	11.49	15.00	18.67	22.48	26.41	30.46	34.61
23	0.46	2.34	4.99	7.81	10.97	14.32	17.81	21.44	25.18	29.03	32.97
24	0.44	2.24	4.68	7.47	10.50	13.69	17.03	20.49	24.06	27.72	31.48
25	0.42	2.15	4.49	7.17	10.06	13.12	16.32	19.62	23.03	26.53	30.11
26	0.40	2.06	4.32	6.88	9.66	12.60	15.66	18.83	22.09	25.44	28.86
27	0.39	1.99	4.15	6.62	9.29	12.11	15.05	18.09	21.22	24.43	27.71
28	0.38	1.92	4.00	6.38	8.95	11.66	14.49	17.41	20.42	23.50	26.65
29	0.36	1.85	3.86	6.15	8.63	11.24	13.97	16.78	19.69	22.64	25.67
30	0.35	1.79	3.73	5.94	8.33	10.86	13.48	16.20	18.99	21.84	24.76
31	0.34	1.73	3.61	5.75	8.06	10.50	13.03	15.65	18.34	21.10	23.91
32	0.33	1.67	3.49	5.56	7.80	10.16	12.61	15.14	17.74	20.40	23.12
33	0.32	1.62	3.39	5.39	7.56	9.84	12.21	14.66	17.18	19.75	22.38
34	0.31	1.58	3.29	5.23	7.33	9.54	11.84	14.21	16.65	19.14	21.68
35	0.30	1.53	3.19	5.08	7.12	9.26	11.49	13.79	16.15	18.57	21.03
36	0.29	1.49	3.10	4.93	6.91	9.00	11.16	13.39	15.68	18.03	20.42
37	0.28	1.45	3.02	4.80	6.72	8.75	10.85	13.02	15.24	17.52	19.84
38	0.28	1.41	2.94	4.67	6.54	8.51	10.56	12.66	14.83	17.04	19.29
39	0.27	1.37	2.86	4.55	6.37	8.29	10.28	12.33	14.43	16.58	18.77
40	0.26	1.34	2.79	4.43	6.07	8.07	10.01	12.01	14.06	16.15	18.28
41	0.26	1.30	2.72	4.32	6.05	7.87	9.76	11.71	13.70	15.74	17.82
42	0.25	1.27	2.65	4.22	5.91	7.68	9.52	11.42	13.36	15.35	17.37
43	0.24	1.24	2.59	4.12	5.76	7.50	9.29	11.15	13.04	14.99	16.95
44	0.24	1.22	2.53	4.02	5.63	7.32	9.08	10.88	12.74	14.63	16.55
45	0.23	1.19	2.47	3.93	5.50	7.16	8.87	10.64	12.44	14.29	16.17
46	0.23	1.16	2.42	3.85	5.38	7.00	8.67	10.40	12.17	13.97	15.81
47	0.22	1.14	2.37	3.76	5.27	6.84	8.48	10.17	11.90	13.66	15.46
48	0.22	1.11	2.32	3.68	5.15	6.70	8.30	9.95	11.64	13.37	15.12
49	0.21	1.09	2.27	3.61	5.05	6.56	8.13	9.74	11.40	13.09	14.80
50	0.21	1.07	2.22	3.53	4.94	6.43	7.96	9.54	11.16	12.82	14.50
60	0.18	0.89	1.85	2.94	4.11	5.34	6.61	7.92	9.26	10.62	12.01
70	0.15	0.76	1.58	2.52	3.52	4.56	5.65	6.77	7.91	9.07	10.26
80	0.13	0.67	1.39	2.20	3.07	3.99	4.93	5.91	6.90	7.92	8.95
90	0.12	0.59	1.23	1.95	2.73	3.54	4.38	5.24	6.13	7.02	7.94
100	0.11	0.53	1.11	1.76	2.45	3.18	3.94	4.71	5.50	6.31	7.13

TABLE 2. Sampling Plans for $P_A = 0.95$

Sample Size n	Acceptance No. c
9	1
21	2
35	3
50	4
67	5
83	6
100	7

These sampling plans can also be obtained from Figure 2.

2. From Figure 1, the sample size corresponding to lot size of 300 is 32.
3. The sampling plan having n close to 32 from Table 2 is

$n = 35 \qquad$ AQL $= 4\%$ (from Figure 2)

$c = 3 \qquad$ LTPD $= 18.1\%$ (from Figure 3, with consumer's risk 10%).

The corresponding sampling plan from Table II-A, MIL-STD-105D, for normal inspection is:

$n = 50 \qquad$ AQL $= 5.36\%$

$c = 5 \qquad$ LTPD $= 17.8\%$

The obvious differences in the two sampling plans are due to two factors: (1) MIL-STD-105D sampling plans are obtained for a range of lot sizes, whereas our method is for a single lot size; (2) MIL-STD-105D has varying α values for the same AQL, whereas in our method α is constant for the set of sampling plans for a given AQL. This method can be extended to different inspection levels and to different switching levels (reduced and tightened). The major advantage is that this technique is easily computer-programmable when compared with the programming of existing published MIL-STD-105D sampling plans.

Similar sampling plans can be obtained by choosing a suitable cost-objective function.

TABLE 1e. Quality Levels for $P_A = 0.95$

n/c	0	1	2	3	4	5	6	7	8	9	10
1	5.00										
2	2.53	22.36									
3	1.70	13.54	36.84								
4	1.27	9.76	24.86	47.29							
5	1.02	7.64	18.93	34.26	54.93						
6	0.85	6.28	15.32	27.13	41.82	60.70					
7	0.73	5.34	12.88	22.53	34.13	47.93	65.18				
8	0.64	4.64	11.11	19.29	28.92	40.03	52.93	68.77			
9	0.57	4.10	9.77	16.88	25.14	34.49	45.04	57.09	71.69		
10	0.51	3.68	8.73	15.00	22.24	30.35	39.34	49.31	60.58	74.11	
11	0.47	3.33	7.88	13.51	19.96	27.12	34.98	43.56	52.99	63.56	76.16
12	0.43	3.05	7.19	12.29	18.10	24.53	31.52	39.09	47.27	56.19	66.13
13	0.39	2.81	6.60	11.27	16.57	22.40	28.70	35.48	42.74	50.53	58.99
14	0.37	2.60	6.11	10.40	15.27	20.61	26.36	32.50	39.04	46.00	53.43
15	0.34	2.42	5.68	9.67	14.17	19.09	24.37	30.00	35.96	42.26	48.92
16	0.32	2.27	5.31	9.03	13.21	17.78	22.67	27.86	33.34	39.10	45.17
17	0.30	2.13	4.99	8.46	12.38	16.64	21.19	26.01	31.08	36.40	41.97
18	0.28	2.01	4.70	7.97	11.64	15.63	19.90	24.40	29.12	34.06	39.22
19	0.27	1.90	4.45	7.53	10.99	14.75	18.75	22.97	27.39	32.01	36.81
20	0.26	1.81	4.22	7.14	10.41	13.96	17.73	21.71	25.87	30.20	34.69
21	0.24	1.72	4.01	6.78	9.88	13.24	16.82	20.57	24.50	28.58	32.81
22	0.23	1.64	3.82	6.46	9.41	12.60	15.99	19.56	23.27	27.13	31.13
23	0.22	1.57	3.65	6.17	8.98	12.02	15.25	18.63	22.16	25.82	29.61
24	0.21	1.50	3.50	5.90	8.59	11.49	14.57	17.80	21.16	24.64	28.24
25	0.20	1.44	3.35	5.66	8.23	11.01	13.95	17.03	20.24	23.56	26.99
26	0.20	1.38	3.22	5.43	7.90	10.56	13.38	16.33	19.40	22.57	25.84
27	0.19	1.33	3.10	5.22	7.59	10.15	12.85	15.68	18.62	21.66	24.79
28	0.18	1.28	2.98	5.03	7.31	9.77	12.37	15.09	17.91	20.82	23.83
29	0.18	1.24	2.88	4.85	7.05	9.42	11.92	14.53	17.25	20.05	22.93
30	0.17	1.20	2.78	4.69	6.81	9.09	11.50	14.02	16.63	19.33	22.11
31	0.17	1.16	2.69	4.53	6.58	8.78	11.11	13.54	16.06	18.66	21.34
32	0.16	1.12	2.60	4.38	6.37	8.50	10.74	13.09	15.53	18.04	20.62
33	0.16	1.09	2.52	4.25	6.17	8.23	10.40	12.68	15.03	17.46	19.95
34	0.15	1.06	2.45	4.12	5.98	7.98	10.08	12.28	14.56	16.91	19.32
35	0.15	1.02	2.38	4.00	5.80	7.74	9.78	11.91	14.12	16.40	18.73
36	0.14	1.00	2.31	3.89	5.64	7.52	9.50	11.57	13.71	15.91	18.18
37	0.14	0.97	2.25	3.78	5.48	7.31	9.23	11.24	13.32	15.46	17.65
38	0.13	0.94	2.19	3.68	5.33	7.11	8.98	10.93	12.95	15.03	17.16
39	0.13	0.92	2.13	3.58	5.19	6.92	8.74	10.64	12.60	14.62	16.69
40	0.13	0.90	2.08	3.49	5.06	6.74	8.51	10.36	12.27	14.24	16.25
41	0.13	0.87	2.02	3.40	4.93	6.57	8.30	10.10	11.96	13.87	15.83
42	0.12	0.85	1.98	3.32	4.81	6.41	8.09	9.85	11.66	13.53	15.44
43	0.12	0.83	1.93	3.24	4.70	6.26	7.90	9.61	11.38	13.20	15.06
44	0.12	0.81	1.88	3.17	4.59	6.11	7.71	9.38	11.11	12.88	14.70
45	0.11	0.80	1.84	3.09	4.48	5.97	7.54	9.17	10.85	12.58	14.36
46	0.11	0.78	1.80	3.02	4.38	5.84	7.37	8.96	10.60	12.30	14.03
47	0.11	0.76	1.76	2.96	4.29	5.71	7.20	8.76	10.37	12.02	13.72
48	0.11	0.75	1.73	2.90	4.20	5.59	7.05	8.57	10.15	11.76	13.42
49	0.10	0.73	1.69	2.84	4.11	5.47	6.90	8.39	9.93	11.51	13.13
50	0.10	0.72	1.66	2.78	4.02	5.36	6.76	8.22	9.72	11.27	12.86
60	0.09	0.60	1.38	2.31	3.34	4.45	5.61	6.81	8.05	9.33	10.64
70	0.07	0.51	1.18	1.98	2.86	3.80	4.79	5.82	6.87	7.96	9.07
80	0.06	0.45	1.03	1.73	2.49	3.32	4.18	5.07	6.00	6.94	7.91
90	0.06	0.40	0.91	1.53	2.21	2.94	3.71	4.50	5.32	6.15	7.01
100	0.05	0.36	0.82	1.38	1.99	2.64	3.33	4.04	4.78	5.53	6.29

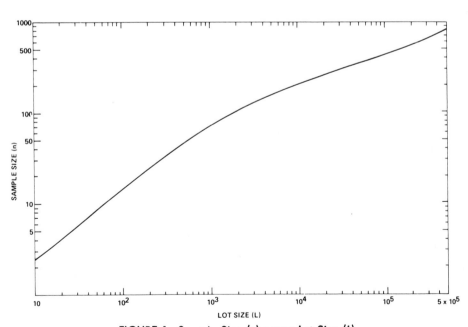

FIGURE 1. Sample Size (n) versus Lot Size (L).

TABLE 1f. Quality Levels for $P_A = 0.975$

n/c	0	1	2	3	4	5	6	7	8	9	10
1	2.50	0.99									
2	1.26	13.46	0.99								
3	0.84	9.42	25.12	0.99							
4	0.63	6.76	19.40	34.66	0.99						
5	0.51	5.27	14.66	28.36	42.28	0.99					
6	0.42	4.32	11.80	22.27	35.87	48.41	0.99				
7	0.36	3.66	9.89	18.40	29.03	42.13	53.42	0.99			
8	0.32	3.18	8.52	15.70	24.48	34.91	47.35	57.57	0.99		
9	0.28	2.81	7.48	13.70	21.20	29.92	39.98	51.75	61.07	0.99	
10	0.25	2.52	6.67	12.15	18.71	26.23	34.75	44.39	55.50	64.06	0.99
11	0.23	2.28	6.01	10.92	16.74	23.37	30.78	39.02	48.22	58.72	66.63
12	0.21	2.08	5.49	9.92	15.16	21.09	27.67	34.87	42.81	51.58	61.52
13	0.19	1.92	5.04	9.08	13.85	19.22	25.13	31.58	38.56	46.18	54.55
14	0.18	1.77	4.65	8.38	12.76	17.66	23.03	28.86	35.12	41.88	49.19
15	0.17	1.65	4.33	7.79	11.82	16.33	21.26	26.58	32.27	38.37	44.89
16	0.16	1.55	4.04	7.26	11.02	15.19	19.75	24.64	29.87	35.42	41.33
17	0.15	1.45	3.79	6.81	10.31	14.21	18.44	22.97	27.81	32.91	38.32
18	0.14	1.37	3.57	6.41	9.69	13.34	17.30	21.53	26.02	30.74	35.73
19	0.13	1.30	3.38	6.04	9.14	12.57	16.29	20.25	24.44	28.86	33.49
20	0.13	1.23	3.20	5.73	8.65	11.89	15.39	19.12	23.05	27.19	31.51
21	0.12	1.17	3.05	5.44	8.22	11.28	14.59	18.10	21.81	25.70	29.77
22	0.12	1.12	2.91	5.19	7.81	10.72	13.86	17.19	20.70	24.39	28.21
23	0.11	1.07	2.77	4.94	7.45	10.22	13.21	16.37	19.70	23.19	26.81
24	0.11	1.02	2.65	4.73	7.13	9.77	12.61	15.63	18.79	22.11	25.55
25	0.10	0.98	2.54	4.53	6.83	9.35	12.06	14.95	17.96	21.12	24.39
26	0.10	0.94	2.44	4.35	6.55	8.97	11.56	14.32	17.21	20.22	23.34
27	0.09	0.90	2.35	4.18	6.30	8.62	11.11	13.75	16.51	19.40	22.38
28	0.09	0.87	2.26	4.03	6.06	8.29	10.68	13.22	15.87	18.64	21.50
29	0.09	0.84	2.18	3.88	5.84	7.99	10.29	12.73	15.28	17.94	20.68
30	0.08	0.82	2.11	3.75	5.63	7.71	9.93	12.28	14.73	17.28	19.92
31	0.08	0.79	2.03	3.63	5.45	7.45	9.59	11.86	14.22	16.68	19.22
32	0.08	0.76	1.97	3.51	5.27	7.21	9.27	11.46	13.74	16.12	18.56
33	0.08	0.74	1.91	3.40	5.10	6.97	8.98	11.09	13.29	15.59	17.95
34	0.07	0.72	1.85	3.29	4.95	6.76	8.70	10.74	12.88	15.10	17.38
35	0.07	0.70	1.80	3.20	4.80	6.55	8.44	10.41	12.48	14.63	16.84
36	0.07	0.68	1.75	3.10	4.67	6.37	8.19	10.11	12.11	14.20	16.34
37	0.07	0.66	1.70	3.02	4.54	6.19	7.96	9.82	11.77	13.79	15.86
38	0.07	0.63	1.65	2.94	4.40	6.02	7.74	9.55	11.44	13.40	15.42
39	0.06	0.62	1.61	2.86	4.29	5.86	7.53	9.29	11.13	13.04	14.99
40	0.06	0.60	1.57	2.79	4.18	5.71	7.34	9.05	10.84	12.69	14.59
41	0.06	0.59	1.53	2.72	4.07	5.56	7.14	8.81	10.56	12.36	14.21
42	0.06	0.58	1.49	2.66	3.97	5.43	6.96	8.59	10.29	12.05	13.85
43	0.06	0.56	1.46	2.58	3.88	5.30	6.80	8.39	10.04	11.75	13.51
44	0.06	0.55	1.43	2.52	3.79	5.17	6.64	8.19	9.80	11.47	13.18
45	0.06	0.54	1.39	2.47	3.70	5.05	6.48	8.00	9.57	11.20	12.87
46	0.06	0.53	1.36	2.41	3.61	4.94	6.34	7.82	9.36	10.95	12.58
47	0.05	0.52	1.34	2.36	3.54	4.83	6.20	7.64	9.15	10.70	12.30
48	0.05	0.50	1.30	2.31	3.47	4.72	6.06	7.48	8.95	10.46	12.03
49	0.05	0.49	1.27	2.26	3.40	4.62	5.94	7.32	8.76	10.24	11.77
50	0.05	0.48	1.25	2.22	3.33	4.52	5.81	7.17	8.57	10.02	11.52
60	0.04	0.40	1.04	1.84	2.75	3.75	4.82	5.94	7.09	8.29	9.52
70	0.04	0.34	0.89	1.57	2.36	3.21	4.11	5.06	6.04	7.06	8.11
80	0.03	0.30	0.78	1.38	2.06	2.79	3.58	4.41	5.27	6.16	7.07
90	0.03	0.27	0.69	1.22	1.82	2.48	3.18	3.91	4.67	5.46	6.26
100	0.03	0.23	0.61	1.09	1.64	2.23	2.86	3.51	4.20	4.90	5.62

BSP for Specified LTPD Values

For the LTPD specification, the Dodge-Romig procedure can be used to obtain sampling plans from Tables 1a through 1f as outlined:

1. For a given lot size and consumer's risk, $\beta(=P_A)$, obtain a set of sampling plans satisfying (11), viz.,

$$\beta = \sum_{r=0}^{c} [(L \times p_t) \, !/r \, ! \times (L \times p_t - r) \, !] \qquad (11)$$
$$[1 - (n/L)]^{L \times p_t - r} \, (n/L)^r ,$$

where L = lot size

p_t = LTPD

n = sample size

β = consumer's risk

c = acceptance number

2. From the set of sampling plans satisfying (11), choose the sampling plan for which the average total amount of inspection (ATI) shown in (12) is minimum,

$$\text{ATI} = n + (L - n)[1 - P_A \text{ at } \bar{p}], \qquad (12)$$

with $\bar{p}$ = process average

P_A = probability of acceptance.

Example

Let LTPD = 10%, $\bar{p} = 4.5\%$, and $\beta = 0.10$.

1. For a given lot size, say, $L = 50$, obtain a set of sampling plans satisfying (11) as follows:

   a. For $\beta = 0.10$, find from Table 1c the quality levels for different c values corresponding to the sample size entry $L \times p_t = 50 \times 0.10 = 5$, viz.,

c	0	1	2	3	4
percent quality levels	36.90	58.39	75.34	88.78	97.91

   b. Calculate the sample sizes for the quality levels in a by the relation $n = L \times$ quality level. Then, the set of sampling plans corresponding to lot size 50 is

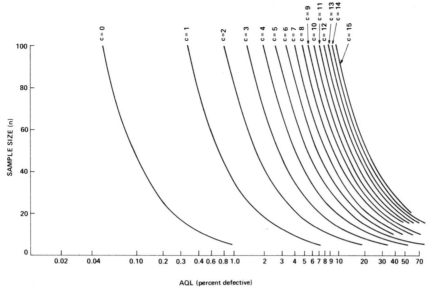

FIGURE 2. Binomial Sampling Plan for Acceptable Quality Level, $P_A = 0.95$.

$n =$ 18 29 38 44 49
$c =$ 0   1   2   3   4

2. For the set of sampling plans in *1, b*, calculate the ATI from (12):

$n =$ 18 29 38 44 49
$c =$ 0   1   2   3   4
ATI = 36 37 41 45 49

The sampling plan that minimizes ATI is (18, 0).

For different lot sizes, this procedure is applied in obtaining Table 3.

TABLE 3. Sampling Plans for $P_A = 0.10$ and LTPD = 0.10

Lot Size	BSP	
	n	c
50	18	0
100	45	2
200	72	4
300	86	5
400	99	6
500	112	7
600	136	9
800	149	10
1,000	161	11

Note: For large lot sizes, the set of binomial sampling plans based on (2) can be obtained from Figure 3 for a given LTPD.

It is to be pointed out that Dodge-Romig derived their sampling plans for a range of lot sizes and process averages through an iterative technique, making adjustments for these ranges. Further, they used a Poisson approximation in (12) for P_A calculations. In our approach, we are using the binomial approximation and only the specified lot size and process average.

BSP for Specified AQL and LTPD

There are industrial situations in which both quality levels AQL and LTPD must be specified simultaneously, as opposed to a single specification of either. Such situations occur most frequently in the in-process manufacturing operations when a product passes through a series of inspection stations. In these instances LTPD values are specified in addition to AQL values to protect the user at the next level of assembly. For these applications binomial sampling plans can be obtained from Tables 1c and 1e with the following procedure.

1. For the given AQL and LTPD values, obtain the operating ratio (O.R.) as
   O.R. = LTPD/AQL.
2. From Table 4, obtain *n* and *c* values for the O.R. obtained in *1*.

Whenever the ratio of sample size to lot size is greater than 0.10, the procedure should be modified as follows.

Obtain sampling plans as solutions to (13) and (14).

$$1 - \alpha = \sum_{r=0}^{c} [(L \times AQL)!/r! \cdot (L \times AQL - r)!] \, (n/L)^r [1 - (n/L)]^{L \times AQL - r} \quad (13)$$

$$\beta = \sum_{r=0}^{c} [(L \times LTPD)!/r!(L \times LTPD - r)!] \cdot (n/L)^r [1 - (n/L)]^{L \times LTPD - r} \quad (14)$$

OPERATING RATIO(BINOMIAL APPROXIMATION)

PRODUCER'S RISK=0.05
CONSUMER'S RISK=0.10

c/n	5	10	15	20	25	30	35	40	45	50
0	36.16	40.20	41.69	42.46	42.93	43.25	43.48	43.65	43.79	43.90
1	7.64	9.16	9.72	10.02	10.20	10.32	10.41	10.47	10.52	10.57
2	3.98	5.15	5.58	5.80	5.94	6.03	6.10	6.15	6.19	6.22
3	2.59	3.68	4.06	4.26	4.38	4.47	4.53	4.57	4.61	4.63
4	1.78	2.90	3.28	3.47	3.58	3.66	3.71	3.76	3.79	3.82
5		2.41	2.79	2.97	3.09	3.16	3.22	3.26	3.29	3.32
6		2.07	2.45	2.64	2.75	2.82	2.88	2.92	2.95	2.98
7		1.79	2.20	2.39	2.50	2.58	2.63	2.67	2.70	2.73
8		1.56	2.00	2.19	2.31	2.39	2.44	2.48	2.51	2.54
9		1.34	1.83	2.04	2.16	2.23	2.29	2.33	2.36	2.39
10			1.69	1.91	2.03	2.11	2.17	2.21	2.24	2.27
11			1.57	1.80	1.92	2.00	2.06	2.10	2.14	2.16
12			1.45	1.70	1.83	1.91	1.97	2.02	2.05	2.08
13			1.34	1.61	1.75	1.84	1.90	1.94	1.97	2.00
14			1.21	1.53	1.68	1.77	1.83	1.87	1.91	1.94
15			1.00	1.46	1.61	1.70	1.77	1.81	1.85	1.88

c/n	5		10		15		20		25		30		35		40		45		50	
	AQL	LTPD	AQL	LTPD	AQL	LTPD	AQL	LTPD	AQL	LTPD	AQL	LTPD	AQL	LTPD	AQL	LTPD	AQL	LTPD	AQL	LTPD
0	1.02	36.90	0.51	20.57	0.34	14.23	0.26	10.87	0.20	8.80	0.17	7.39	0.15	6.37	0.13	5.59	0.11	4.99	0.10	4.50
1	7.64	58.39	3.68	33.68	2.42	23.56	1.81	18.10	1.44	14.69	1.20	12.36	1.02	10.66	0.90	9.38	0.80	8.37	0.72	7.56
2	18.93	75.34	8.73	44.96	5.68	31.73	4.22	24.48	3.35	19.91	2.78	16.78	2.38	14.50	2.08	12.76	1.84	11.40	1.66	10.30
3	34.26	88.78	15.00	55.17	9.67	39.28	7.14	30.42	5.66	24.80	4.69	20.93	4.00	18.10	3.49	15.94	3.09	14.25	2.78	12.88
4	54.93	97.91	22.24	64.58	14.17	46.40	10.41	36.07	8.23	29.47	6.81	24.90	5.80	21.55	5.06	19.00	4.48	16.98	4.02	15.35
5			30.35	73.27	19.09	53.17	13.96	41.49	11.01	33.97	9.09	28.74	7.74	24.90	6.74	21.96	5.97	19.64	5.36	17.76
6			39.34	81.24	24.37	59.65	17.73	46.73	13.95	38.33	11.50	32.47	9.78	28.15	8.51	24.85	7.54	22.23	6.76	20.11
7			49.31	88.42	30.00	65.85	21.71	51.80	17.03	42.58	14.02	36.11	11.91	31.34	10.36	27.67	9.17	24.77	8.22	22.42
8			60.58	94.55	35.96	71.78	25.87	56.73	20.24	46.73	16.63	39.68	14.12	34.46	12.27	30.45	10.85	27.27	9.72	24.69
9			74.11	98.95	42.26	77.44	30.20	61.52	23.56	50.80	19.33	43.19	16.40	37.54	14.24	33.18	12.58	29.73	11.27	26.92
10					48.92	82.80	34.69	66.18	26.99	54.77	22.11	46.63	18.73	40.56	16.25	35.88	14.36	32.16	12.86	29.13
11					56.02	87.82	39.36	70.71	30.51	58.67	24.95	50.01	21.12	43.54	18.31	38.53	16.17	34.55	14.47	31.31
12					63.66	92.41	44.20	75.09	34.14	62.49	27.87	53.34	23.56	46.48	20.41	41.16	18.01	36.92	16.12	33.47
13					72.06	96.40	49.22	79.33	37.86	66.23	30.85	56.62	26.05	49.38	22.55	43.75	19.89	39.27	17.79	35.60
14					81.90	99.30	54.44	83.41	41.68	69.89	33.89	59.85	28.58	52.25	24.73	46.32	21.80	41.50	19.49	37.72
15					0.00	0.00	59.90	87.31	45.61	73.47	36.99	63.03	31.17	55.08	26.94	48.86	23.73	43.88	21.21	39.82

TABLE 4. Binomial Sampling Plan for Operating Ratios with Given Acceptable Quality Level and Lot Tolerance Percent Defective Specifications (Continued)

OPERATING RATIO(BINOMIAL APPROXIMATION)

PRODUCER'S RISK=0.05
CONSUMER'S RISK=0.10

c/n	60	70	80	90	100
0	44.06	44.18	44.26	44.33	44.39
1	10.63	10.67	10.71	10.73	10.75
2	6.27	6.30	6.33	6.35	6.36
3	4.68	4.71	4.73	4.75	4.76
4	3.86	3.88	3.91	3.92	3.94
5	3.35	3.38	3.40	3.42	3.43
6	3.01	3.04	3.06	3.08	3.09
7	2.77	2.79	2.81	2.83	2.84
8	2.58	2.60	2.62	2.64	2.65
9	2.43	2.45	2.47	2.49	2.50
10	2.30	2.33	2.35	2.37	2.38
11	2.20	2.23	2.25	2.27	2.28
12	2.12	2.14	2.17	2.18	2.20
13	2.04	2.07	2.09	2.11	2.12
14	1.98	2.01	2.03	2.04	2.06
15	1.92	1.95	1.97	1.99	2.00

c/n	60		70		80		90		100	
	AQL	LTPD	AQL	LTPD	AQL	LTPD	AQL	LTPD	AQL	LTPD
0	0.09	3.76	0.07	3.24	0.06	2.84	0.06	2.53	0.05	2.28
1	0.60	6.33	0.51	5.44	0.45	4.78	0.40	4.25	0.36	3.83
2	1.38	8.63	1.18	7.42	1.03	6.52	0.91	5.81	0.82	5.23
3	2.31	10.80	1.98	9.30	1.73	8.16	1.53	7.27	1.38	6.56
4	3.34	12.88	2.86	11.10	2.49	9.74	2.21	8.69	1.99	7.83
5	4.45	14.91	3.80	12.85	3.32	11.28	2.94	10.06	2.64	9.08
6	5.61	16.89	4.79	14.56	4.18	12.79	3.71	11.41	3.33	10.29
7	6.81	18.84	5.82	16.24	5.07	14.28	4.50	12.73	4.04	11.49
8	8.05	20.75	6.87	17.90	6.00	15.74	5.32	14.04	4.78	12.67
9	9.33	22.64	7.96	19.54	6.94	17.18	6.15	15.33	5.53	13.84
10	10.64	24.51	9.07	21.15	7.91	18.60	7.01	16.60	6.29	14.99
11	11.97	26.36	10.20	22.75	8.89	20.02	7.88	17.86	7.07	16.13
12	13.32	28.19	11.35	24.34	9.89	21.42	8.76	19.12	7.86	17.26
13	14.69	30.00	12.52	25.91	10.90	22.80	9.66	20.36	8.67	18.39
14	16.09	31.80	13.70	27.47	11.93	24.18	10.56	21.59	9.48	19.50
15	17.50	33.58	14.89	29.02	12.97	25.55	11.48	22.82	10.30	20.61

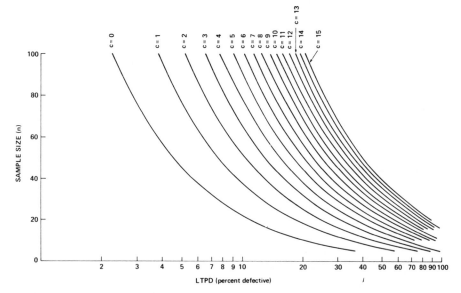

FIGURE 3. Binomial Sampling Plan for Lot Tolerance Percent Defective, $P_A = 0.10$.

Then, Tables 1a through 1f can be used to obtain the appropriate sampling plans simply by using n for $L \times$ AQL (and n for LTPD) and p for n/L. This procedure is illustrated in the following example.

Example

Suppose AQL $= 5\%$, $\alpha = 0.05$

LTPD $= 14\%$, $\beta = 0.10$

Lot size $(L) = 500$

1. O.R. $= 14/5 = 2.8$

2. From Table 4, the sampling plan corresponding to 2.8 is

$n = 80$

$c = 7$.

3. Since $n/L = 0.16$ is greater than 0.10, a sampling plan is obtained as follows:

   a. $L \times$ AQL $= 500 \times 0.05 = 25 \sim n$ (in Table 1e)
   $L \times$ LTPD $= 500 \times 0.14 = 70 \sim n$ (in Table 1c)

   b. Using Tables 1a through 1f for the n values in a, we obtain the quality levels in Table 5 around the c value in step 2.

   c. From Table 5, $c = 6$ is selected because it corresponds to approximately the same quality level for both $n = 25$ and $n = 70$.

   d. Sample size (n) is obtained from the relation $n/L =$ quality level. Table 6 is generated for the appropriate quality levels in c, with the

associated α and β risks. Depending upon the emphasis on α or β, the appropriate sampling plan can be chosen from Table 6.

The adjusted BSP from Table 6 provides less sampling inspection cost when compared with the sampling plan $n = 80$ and $c = 7$, obtained by the unadjusted binomial approximation of Table 4, or when compared with the sampling plan $n = 93$ and $c = 8$, obtained by the Poisson approximation method [2].

AQL Plans Based on Cost Break-Even Percentage Defectives

Enell [1] has given a criterion for selecting an AQL plan by balancing the cost of finding and cor-

TABLE 5. Quality Levels for $n = 25$ and $n = 70$

	Quality Levels For	
c	n = 25	n = 70
5	11.01	12.85
6	13.95	14.56
7	17.03	16.24
8	20.24	17.90

TABLE 6. Adjusted Binomial Sampling Plan

AQL $= 5\%$ $\alpha = 0.05$

LTPD $= 14\%$ $\beta = 0.10$

Quality level	BSP				Exact Value	
	n	c	α	β	α	β
13.95	70	6	0.051	0.124	0.046	0.106
14.56	73	6	0.062	0.098	0.057	0.082

recting a defective against the loss incurred if a defective escapes through inspection. The break-even equation given by Enell is

$$A = (I/p') + C \qquad (15)$$

where

A = unit cost of acceptance (damage done by a defective piece which slips through inspection)

I = unit cost of inspection

C = cost of reworking or replacing a defective once found

p' = unknown fraction defective in the lot.

The break-even percent defective (p_b) is given by

$$p_b = I/(A - C). \qquad (16)$$

From p_b, a MIL-STD-105D sampling plan is chosen so that the OC curve of the plan has $P_A = 0.5$ at p_b. With the binomial approximation for the calculation of P_A values, AQL sampling plans can be obtained by the following procedure.

1. From the cost estimates, obtain p_b with (16).
2. Using Figure 4, obtain a set of sampling plans satisfying P_A at $p_b = 0.5$.
3. From Figure 1, obtain the sample size for the given lot size.
4. Choose the sampling plan from the set in step 2 so that the sample size in step 3 is close to the values in step 2.

Example

Given: $L = 200$

TABLE 7. Sampling Plans for Indifference Quality

n	c
4	0
11	1
18	2
24	3
30	4

$p_b = 0.15$

P_A at $p_b = 0.50.$

The set of sampling plans from Figure 4 is given in Table 7.

From Figure 1 for lot size 200, the sample size is 24. Therefore, the sampling plan satisfying the requirements is $n = 24$ and $c = 3$.

With MIL-STD-105D tables used for break-even percent defective the following sampling plans are the choices:

$n = 32$ $n = 32$

$c = 3$ or $c = 5$

P_A at $p_b = 0.27$ P_A at $p_b = 0.65$

These sampling plans indicate that the binomial approximation results in a reduction of the sample size and maintains $P_A = 0.5$ at p_b. The BSP obtained here again corresponds to a single value of the lot size, as opposed to a range in the other systems of sampling plans.

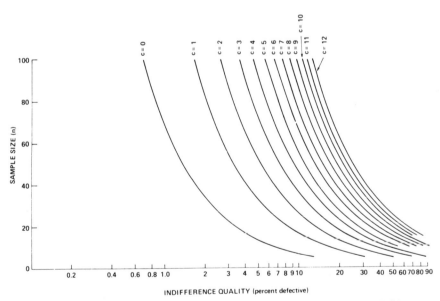

FIGURE 4. Binomial Sampling Plan for Indifference Quality, $P_A = 0.50.$

Confidence Intervals for Population-Proportion Defectives

From the sampling inspection results, it is frequently important to determine an interval that will have a high probability of including the population value. These confidence intervals can be constructed by means of the binomial approximation. The approximate confidence intervals are obtained as a solution of (17) and (18):

$$\sum_{r=0}^{D-1} \binom{n}{r} p_L^r (1 - p_L)^{n-r} = 1 - \alpha/2, \quad (17)$$

$$\sum_{r=0}^{D} \binom{n}{r} p_U^r (1 - p_U)^{n-r} = \alpha/2, \quad (18)$$

where p_L = lower $100 (1 - \alpha)\%$ confidence limit

p_U = upper $100 (1 - \alpha)\%$ confidence limit

α = confidence coefficient

n = sample size

D = observed number of defectives in the sample.

The procedure to obtain the confidence interval from Tables 1a through 1f is as follows:

1. For the given n, D, and α, use the table (of Tables 1a through 1f) for $P_A = 1 - \alpha/2$.
   For the values of $D - 1$ and n, read from the table the value of p_L.
2. Using the table for $P_A = \alpha/2$, obtain the value of p_U for D and n values.
3. A $100 (1 - \alpha)\%$ confidence interval is (p_L, p_U).

Example

Suppose sample size $n = 50$, and observed number of defectives in the sample, $D = 3$.

To find a 95% confidence interval for the population percent defective:

1. From Table 1f, with $n = 50$ and $D = 2$, p_L is 1.25%.
2. From Table 1a, with $n = 50$ and $D = 3$, p_U is 16.56%.
3. Thus, a 95% confidence interval for the population percent defective is (1.25%, 16.56%) when a sample of size 50 contains three defectives.

References

Periodicals:

1. Enell, J. W., "What Sampling Plans Should I Choose?" *Industrial Quality Control*, Vol. 10, No. 6, May 1954, pp. 96–100.
2. Grubbs, F. E., "On Designing Single Sampling Inspection Plans," *Annals of Mathematical Statistics*, Vol. XX, 1949, pp. 242–256.
3. Hamaker, H. C., "The Theory of Sampling Inspection Plans," *Philips Technical Review*, Vol. 11, 1950, pp. 260–270.

Books:

4. Dodge, H. F., and Romig, H. G., *Sampling Inspection Tables*, John Wiley and Sons, Inc., New York, 1959.
5. Johnson, N. L., and Katz, S., *Distributions in Statistics (Discrete Distributions*, Vol. I), Houghton Miflin Co., New York, 1969.

Company and Government Reports:

6. "Sampling Procedures and Tables for Inspection by Attributes," MIL-STD-105D, April 1963, Department of Defense, Washington, D. C.

CHAPTER 3

APPLICATIONS

Presented at the CoG/SME Composites in Manufacturing 6
Conference, January 1987

Quality Control Testing for Woven Glass Prepregs

Chia-Chieh Chen, Mitch Mehlman and George Thomas
FMC Corporation

INTRODUCTION

Quality control of composite components begins with quality control of the constituent materials. The initial quality control process is crucial because all subsequent operations are dependant on the quality of the raw materials.

Preimpregnated materials or prepregs consist of reinforcing fibers and resin with catalyst. Their properties are dependent on storage time and conditions.

In general, the quality of the prepreg is determined by the following: Fiber orientation and quality, uniformity, wetting of the fibers, resin content, volatiles, resin formulation and viscosity, moisture content, tack and cure state of the resin. Any of these areas can cause major effects on the properties of the finished part.

High performance liquid chromatography (HPLC), Fourier transform infrared spectroscopy (FTIR), differential scanning calorimetry (DSC) and other analytical methods have been proven to provide adequate resin information for prepreg quality control (ref. 6,7,9). Since prepregs are used for structural components, the cured laminate properties are important. Several chemical, thermal and mechanical tests were evaluated and compared in order to select meaningful quality control methods.

Woven roving fiber reinforced prepregs are different from unidirectional tapes in that the fiber bundles are woven together. The properties and processing characteristics of woven glass composites are different from laminates made with unidirectional tape.

The objective of this study was to evaluate different test methods for woven glass prepregs. The sensitivity to resin cure, accuracy, data interpretation, cost and limitations of several test methods were compared. Tests were designed such that the effects of prepreg storage, processing and curing on the laminate quality were also studied.

The results of this study are not limited to woven glass prepregs. A similar approach may be used with other materials. The quality control of composites requires careful control of all the manufacturing operations shown in the flow chart below:

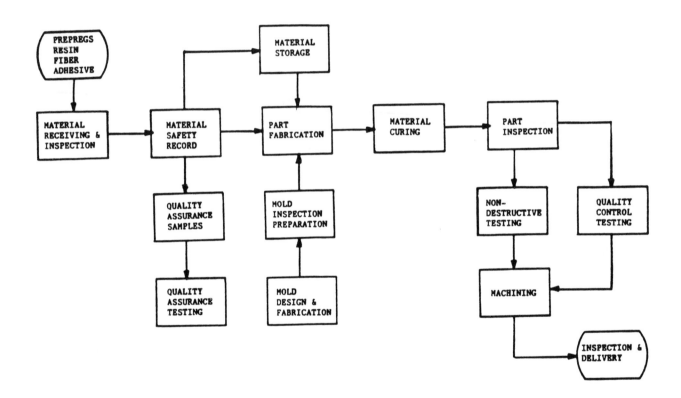

EXPERIMENTAL

In order to evaluate the accuracy, reproducibility, sensitivity and value of different test methods, both cured and uncured materials were tested. Three sets of 4-ply laminates were made using different combinations of materials, prepreg storage conditions, molding methods and curing conditions.

Prepreg Material

A typical, 250°F curing E-glass woven roving/epoxy (DGEBA/DICY) prepreg and an S-2 glass woven roving/polyester (elastomer modified isophthalic DAP monomer based/BPO) material were chosen for evaluation. Properties of the prepregs are listed below.

Property	Epoxy/E-glass	Polyester/S-2^R Glass
Resin content (%)	33.4	30.0
Volatiles (%)	1.5	2.7
Flow (%)	13.0 (50 psi)	3.7 (15 psi)
Gel time @275°F (min.)	4.8	1.05

Laminate Fabrication & Sample Preparation

All vacuum bagged laminates were made with 4 plies of prepreg cut from the same roll of material. The prepreg was placed in a re-usable silicone rubber bag with a standard vacuum port and clamp frame. A thermocouple was inserted in the center of the laminate. Vacuum was applied and the laminates were cured in an circulating air oven. Pressure molded laminates were placed between the platens of a 20 ton hydraulic press and pressure was applied during curing.

Mechanical, thermal and chemical testing was performed on the cured laminates. The test samples were made with standard templates. Extreme care was taken to align the fibers in the test direction. Acetone extraction of the resins was performed for preparation of several FTIR and DSC samples. The same tests were performed on cured laminates without extraction.

Effects of Storage Conditions Prepreg Properties

To prevent undesired curing, prepregs are stored in a sub-ambient temperature environment (freezer) and returned to room temperature before use. After each use, the material is sent back to cold storage. This type of temperature cycling has a cumulative effect on the prepreg's properties. Since the effects of storage depend on the resin chemistry, it was decided to evaluate the typical epoxy/glass and the polyester/glass prepregs under different storage conditions. Tests were performed on the cured laminates and the prepregs.

	Epoxy/E-glass	Polyester/S-2^R Glass
Storage Conditions		
Baseline	As-received	As-received
Room Temp.	2 weeks	2 weeks
100°F	3 days	3 days
100°F	2 weeks	2 weeks
Molding condition	260°F, 1.5 hour	250°F, 2 hour + 275°F, 2 hour

Effect of Molding Process Conditions

Vacuum bag and autoclave molding are typical methods used to compact the laminate during the cure cycle. Laminates can also be cured in a press. In this study, vacuum bagging and press curing were used to make 4-ply laminates. The effects of bleeder ply and pressure were compared for these two processes. The following molding conditions were used:

Molding Process

Vacuum with no bleeder
Vacuum with bleeder
Press with no bleeder
Press with bleeder

Effects of Cure Temperature and Time

An epoxy/glass panel was made with the recommended cure temperature and a second panel was cured at a slightly lower temperature. Several samples taken from the first panel were then aged in an oven before testing.

Normal Cure	Aged	Low Temp. Cure
30 min. @ 150°F	normal cure	30 min. @ 190°F
+	+	+
30 min. @ 200°F	72 hr. @350°F	30 min. @ 230°F
+		
1 hr. @ 260°F		

Tests Performed

Three to six samples were tested for all mechanical and chemical tests.

. Tensile Tests: ASTM D638, ASTM D3039
. Compressive Test: ASTM D695
. Short Beam Shear Test: ASTM D2344
. Flexural Test, 3 point: ASTM D790
. Microscopic Examination
. Infrared Spectroscopy (FTIR)
. Dynamic Mechanical Analysis (DMA)
. Hardness, Barcol: ASTM D2583
. Density: ASTM D792
. Resin Content: ASTM D2584
. Void Content: ASTM D2734

RESULTS

Test results are listed in Tables I through V. In order to evaluate each test method, the data from all the tables were summarized. It should be noted that the charts represent data collected from samples prepared under different conditions. They were used to study the significance of the results rather than to evaluate the effects of processes and conditions.

Tensile Test: The standard deviation of each data point is reasonably low.
The test was not very sensitive to changes in resin cure, molding process
or storage conditions. The aged epoxy sample (4) was clearly degraded and
showed reduced strength.

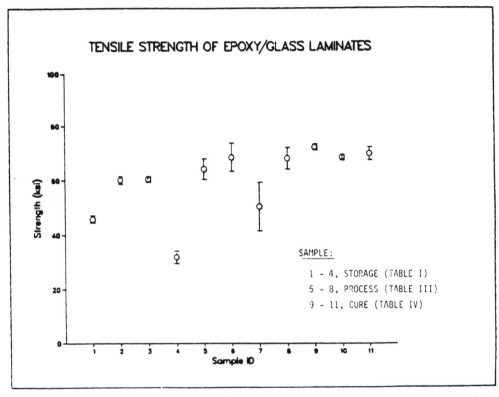

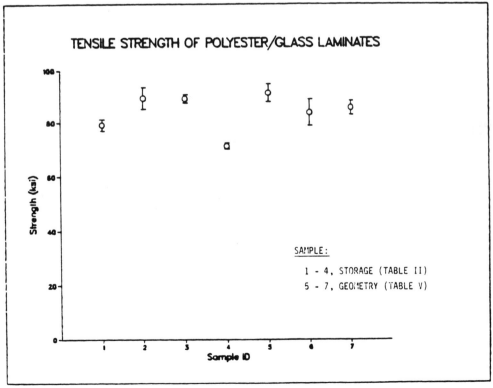

Compressive Test: Test results show higher standard deviation than that of the tensile test. Though the compressive strength is a matrix dominated property, it was not very sensitive to most of the experimental conditions. Again, the aged epoxy sample (4) shows a drastic strength reduction due to matrix degradation.

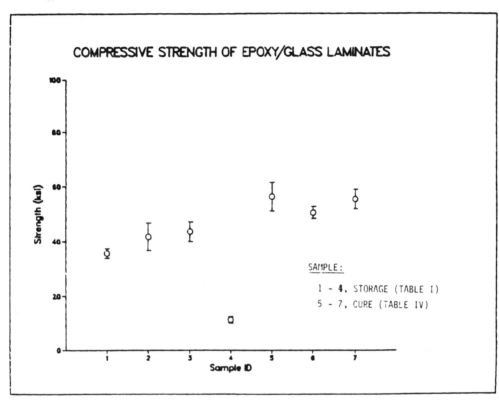

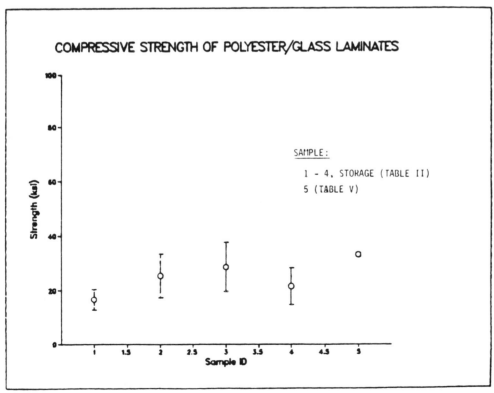

Short Beam Shear: Similar to the other mechanical tests, there were no obvious trends due to resin cure, molding process or prepreg storage conditions. Shear strength reduction was also evident in the aged epoxy sample (4).

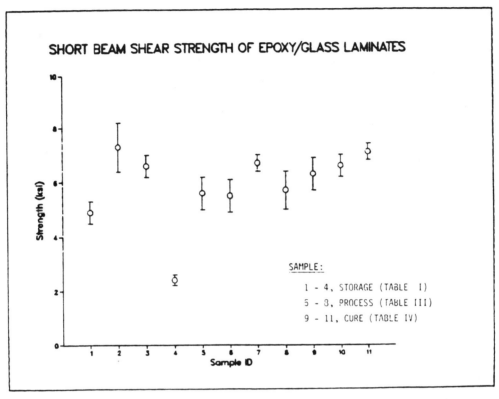

SHORT BEAM SHEAR STRENGTH OF EPOXY/GLASS LAMINATES

SAMPLE:

1 - 4, STORAGE (TABLE I)
5 - 8, PROCESS (TABLE III)
9 - 11, CURE (TABLE IV)

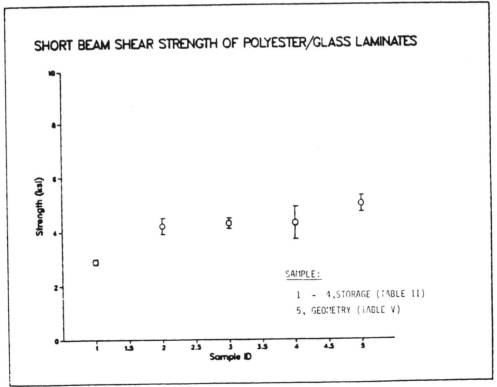

SHORT BEAM SHEAR STRENGTH OF POLYESTER/GLASS LAMINATES

SAMPLE:

1 - 4, STORAGE (TABLE II)
5, GEOMETRY (TABLE V)

Flexural Test: The three point bending test was used. The results showed the importance of sample size. The strength is in the same range as the tensile strength.

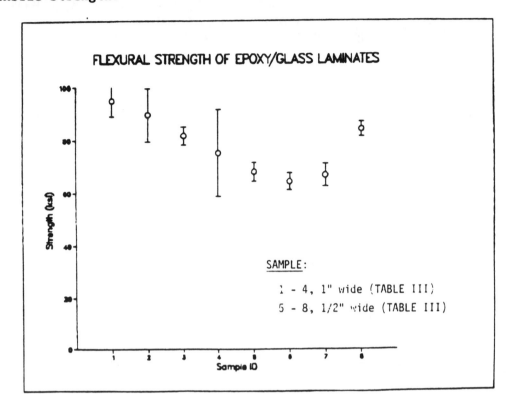

DMA Analysis: This method showed the best sensitivity to resin cure and prepreg storage. The chart below clearly showed the shift of the glass transition temperature (Tg) as a function of resin cure (Table IV).

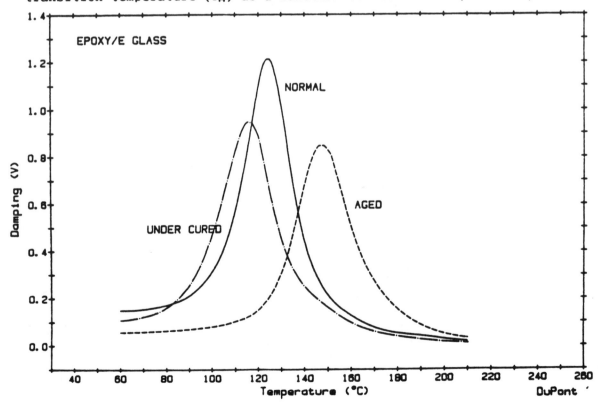

DMA Analysis: Laminates made with prepregs stored under different conditions showed differences in their Tg. The epoxy laminate made with two week oven aged material was clearly degraded and showed a decrease in Tg.

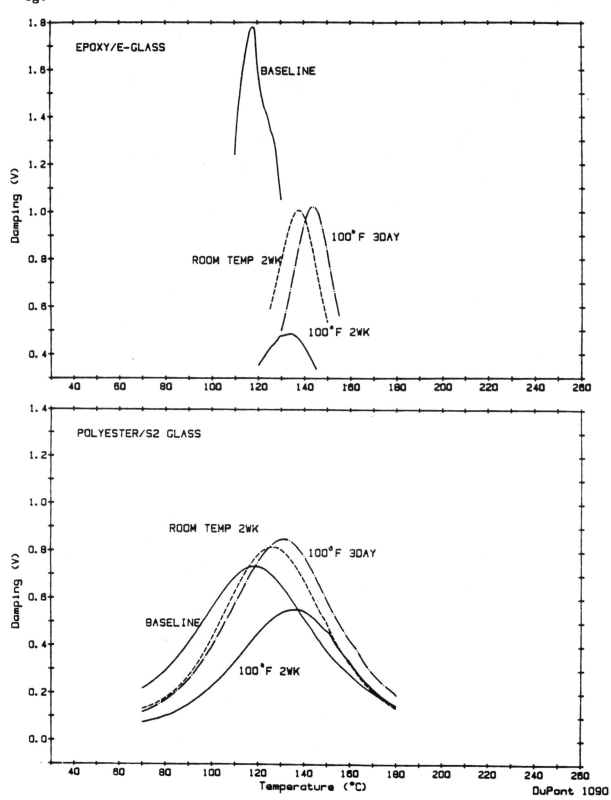

Microscopic Examination: At 40X magnification, the size and distribution of voids in the laminates are compared. Sample A is the baseline epoxy panel and sample B is the laminate made with three day oven aged prepreg (Table I). The unusually large voids resulted in the low mechanical strength of the baseline laminate.

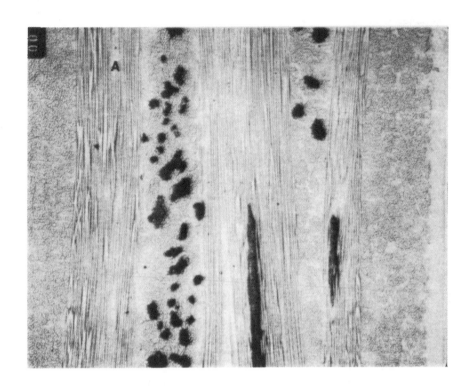

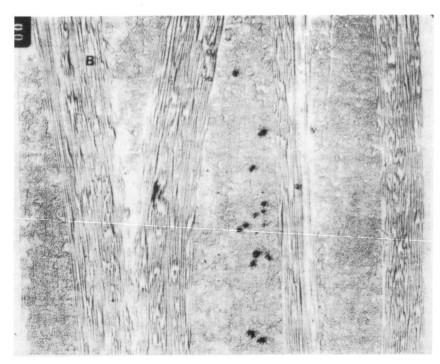

Microscopic Examination: At higher magnification (200X), the size, shape and even depth of the voids can be compared. These are the same laminates as shown on the previous page.

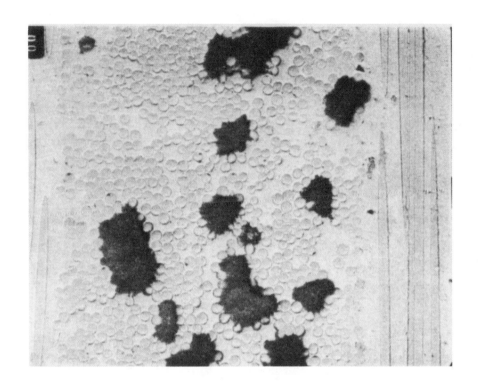

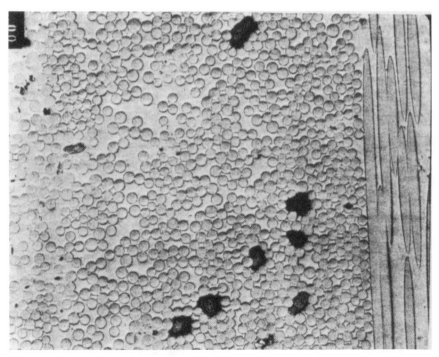

Microscopic Examination: The sample (40X and 200X) was taken from the polyester laminate made with three day oven aged prepreg. The laminate has many more voids than the epoxy laminate, however the calculated result (Table II) was less. At high magnification (200X), contaminant (arrow) and cracks can be seen.

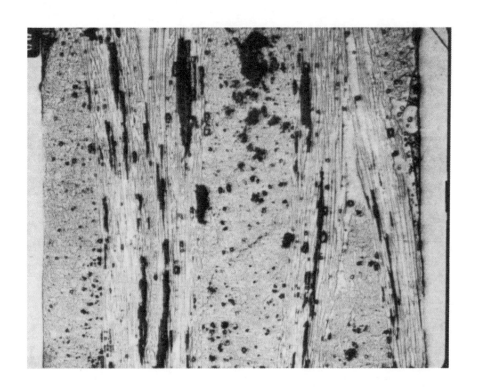

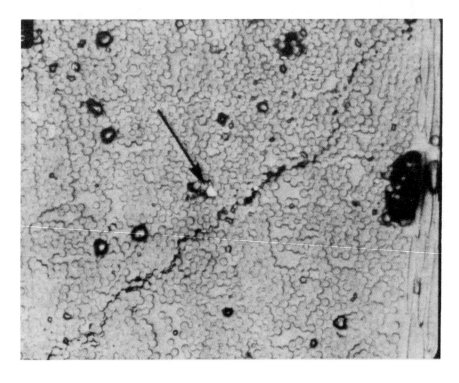

FTIR Analysis: IR analysis was performed on prepregs as well as the cured laminates. The charts shown here are taken from the uncured prepregs after different storage conditions. The peaks at 2180cm^{-1} (Nitride) and 1510cm^{-1} seem to be quantitative. Previous studies (reference 1, 10) have shown the sensitivity and reproducibility of this technique for prepreg analysis. However, when different location from the same cured panel were checked with FTIR, variations of some peaks were found. The method can be quantitative but detailed baseline information and data analysis may be required.

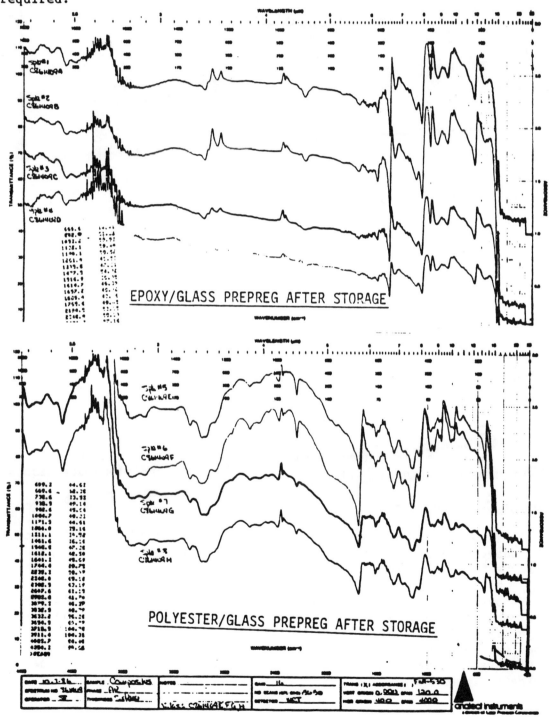

EPOXY/GLASS PREPREG AFTER STORAGE

POLYESTER/GLASS PREPREG AFTER STORAGE

Hardness, Density, Void Content, Resin Content: These tests can be used to determine the general properties of the laminate and should be used in combination with other tests. Except for a few points, most void contents ranged from 1 to 4 %. It should be noted that the calculation of void content is very sensitive to density measurement.

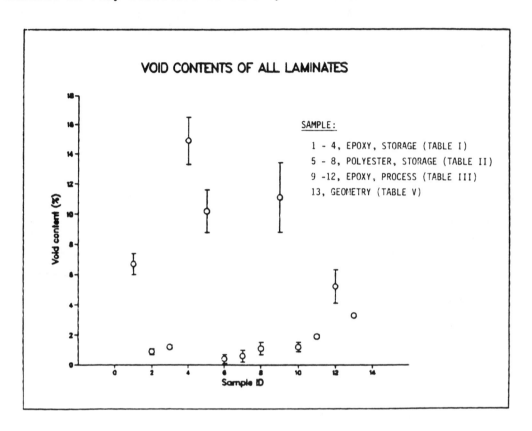

DISCUSSION

The discussion is concentrated on the results from each data table. The effect of storage conditions, process and molding conditions are discussed.

Storage Conditions

In Table I, the effect of storage on epoxy prepreg is clearly identified by the change in glass transition temperature (Tg) as detected by DMA. The increase in Tg indicates an increase in the degree of cure of the resin.

The low mechanical properties of the epoxy/glass baseline panel were attributed to the unusually high void contents of the panel. The data, when compared with data of the other tables, indicated the quality of the baseline panel was questionable. The epoxy prepreg, after storage in the oven for 2 weeks, was hardened and was difficult to mold. This resulted in the poor laminate properties.

Table II shows the properties of polyester/glass laminates after different prepreg storage conditions. The best correlation to resin cure is shown by the Tg shift as measured by DMA. The low mechanical properties of the baseline panel were attributed to the high void content.

Though the calculated void content of the polyester panels was not much higher than those of epoxy panels, microscopic examination revealed higher void content in all polyester panels. This may explain in part the very low compressive and SBS strength of these panels.

The polyester/glass system seemed to show better property retention than the epoxy system after the same temperature exposure. Compression and SBS properties of composites are generally matrix dominated and it was expected that epoxies would show higher properties. Tensile strength of composites is fiber dominated; the polyester panels were made of S-2[R] Glass which resulted in higher tensile strength than the E-glass laminates.

Process Conditions

Vacuum bagging and press molding are compared in Table III for epoxy/glass prepregs. The unusually low tensile strength of the vacuum bagged, no bleeder samples was attributed to its low fiber content and mis-aligned fiber in the samples. Though only 4 plies of prepreg were used for the panel, fiber orientation could not be controlled as well as with unidirectional tape. Examination of the samples showed slight fiber mis-alignment in the test direction. On the other hand, the SBS strength of the panel is the highest among all four panels; this shows the importance of the resin matrix in shear failure.

Cure Conditions

In Table IV, the panel molded at lower than suggested cure temperature showed higher strength than the panel molded at the suggested cure temperature. Again, the mechanical properties seem to correlate well with the void content but not with the resin cure. DMA measurements identified the Tg of each panel and indicated the lower Tg of the low temperature cured panel. Heat aging of the panel resulted in higher tensile, SBS and compressive strength as well as hardness. Apparently, these short term mechanical tests may not be able to detect the effect of aging.

Specimen Geometry

Tensile testing of unidirectional composites (ASTM D3039) requires end tabs which can be expensive to fabricate. Dog-bone shaped samples (ASTM D638) are generally used for testing unfilled and short fiber reinforced plastics. The two methods were compared.

Table V shows the comparable and consistent strength of the dog-bone samples. The standard 1" wide ASTM D3039 samples had the highest tensile strength with medium standard deviation. The same type of samples with narrow gage section yielded lower values. The dog-bone samples showed slightly lower but more consistent tensile test results. Examination of the tested samples confirmed the proper tensile failure; this indicates that dog-bone samples are suitable for testing woven prepregs.

SUMMARY

Mechanical tests, though standard for many quality control applications, are difficult to control. The standard deviation of the data can range from 1% to over 10%. The tests are dependent on many material, processing and testing parameters. If mechanical tests are specified, extreme care must be taken to ensure proper control of the samples and the test. Other tests are needed to confirm the results and sample examination is important after the tests.

For woven glass composites, ASTM D638 (dog-bone) specimens showed comparable results to to ASTM D3039 (end tabbed) specimens.

Storage and temperature cycling of prepregs has major effects on the dynamic properties of the laminates as detected by DMA. Short beam shear and compressive tests are resin dominated and may be used to check the quality of resin/fiber interface. Tensile tests, on the other hand, are less sensitive to the cure of the resin.

Fiber content, void content and hardness are simple tests from which useful information can be derived. Reproducibility and accuracy are major concerns.

DMA seems to show a lot of promise as a method to determine the resin cure state. This test checks both the resin and the fiber as a unit. It has consistently differentiated the cured composite laminates where mechanical tests did not. Modulus and damping information can also be derived from the test. Excellent reproducibility has also been demonstrated.

FTIR is potentially a quantitative method for checking prepreg quality. however, sample preparation and data analysis may require some effort for practical applications.

Quality control of woven glass prepregs and laminates may be accomplished by combination of chemical, thermal and mechanical tests. Quality control testing depends on the material, sample preparation, testing procedures and data interpretation.

TABLE I

EFFECT OF STORAGE CONDITIONS ON EPOXY PREPREGS

PROPERTY	BASELINE	ROOM TEMP 2 WK	100°F 72 H	100°F 2 WK
Glass transition Temperature (°C)	119	138	144.5	134.5
Tensile Strength (Ksi)				
Avg.	45.8	60.2	60.6	31.8
Std Dev.	1.3	1.4	1.0	2.3
SBS strength (Ksi)				
Avg.	4.9	7.3	6.6	2.4
Std Dev.	0.4	0.9	0.4	0.2
Compression (Ksi)				
Avg.	35.6	41.7	43.6	11.3
Std Dev.	1.7	5.1	3.6	1.2
Resin content (%)				
Avg.	32.8	32.2	32.1	31.5
Std Dev.	0.5	0.8	0.2	0.3
Void Content (%)				
Avg.	6.1	0.9	1.2	14.9
Std Dev.	0.7	0.2	0.1	1.6
Density (g/cm^3)				
Avg.	1.81	1.92	1.92	1.66
Std Dev.	0.02	0.01	0.01	0.01
Hardness, Barcol	62	59	64	58

TABLE II

EFFECT OF STORAGE CONDITIONS ON POLYESTER PREPREGS

PROPERTY	BASELINE	RM TEMP 2 WK	100°F 72 H	100°F 2 WK
Glass transition Temperature (°C)	119	127	131.5	136.5
Tensile Strength (Ksi)				
Avg.	79.2	89.3	89.1	71.3
Std Dev.	2.1	4.2	1.6	1.1
SBS strength (Ksi)				
Avg.	2.9	4.2	4.3	4.3
Std Dev.	0.1	0.3	0.2	0.6
Compression (Ksi)				
Avg.	16.7	25.6	28.5	21.4
Std Dev.	3.8	8.0	9.1	6.7
Resin content (%)				
Avg.	28.7	28.9	28.7	28.5
Std Dev.	0.1	0.2	0.3	0.2
Void Content (%)				
Avg.	10.2	0.4	0.6	1.1
Std Dev.	1.4	0.3	0.4	0.4
Density (g/cm^3)				
Avg.	1.70	1.89	1.89	1.92
Std Dev.	0.03	0.01	0.01	0.01
Hardness, Barcol	56	60	60	63

TABLE III

EFFECT OF MOLDING PROCESS ON LAMINATE PROPERTIES

PROPERTY	PRESS WITH NO BLEEDER	PRESS WITH BLEEDER	VACUUM WITH NO BLEEDER	VACUUM WITH BLEEDER
Tensile Strength (Ksi)				
Avg.	64.2	68.5	50.4	68.1
Std Dev.	3.8	5.2	9.0	3.9
Short beam shear (Ksi)				
Avg.	5.6	5.5	6.7	5.7
Std Dev.	0.6	0.6	0.3	0.7
Flexural strength (Ksi)				
1" wide sample				
Avg.	94.9	89.6	81.8	75.2
Std Dev.	5.7	10.0	3.4	16.4
1/2" wide sample				
Avg.	68.1	64.4	66.9	84.2
Std Dev.	3.6	3.3	4.2	2.8
Flexural modulus (Msi)				
1" wide sample				
Avg.	3.6	2.8	2.9	3.0
Std Dev.	0.2	0.2	0.2	0.2
1/2" wide sample				
Avg.	2.8	2.9	3.0	3.3
Std Dev.	0	0.1	0.4	0.2
Density (g/cm^3)	1.7	2.0	1.9	1.9
Resin content (%)				
Avg.	31.6	25.1	32.5	27.4
Std Dev.	1.6	0.9	0.2	1.2
Void content (%)				
Avg.	11.1	1.2	1.9	5.2
Std Dev.	2.3	0.3	0.1	1.1

TABLE IV

EFFECT OF RESIN CURE ON LAMINATE PROPERTIES

PROPERTY	LOW TEMP. CURED	NORMAL CURE	HEAT AGED
Tensile strength (Ksi)			
Avg.	72.2	68.5	69.8
Std Dev.	1.1	1.0	2.4
Tensile modulus (Msi)			
Avg.	3.3	3.4	3.2
Std Dev.	0.1	0.1	0.1
Short Beam Shear (Ksi)			
Avg.	6.3	6.6	7.1
Std Dev.	0.6	0.4	0.3
Compressive strength (Ksi)			
Avg.	56.3	50.4	55.3
Std Dev. (psi)	5.2	2.2	3.6
Glass transition (°C)			
by DMA	116	121	148
Density (g/cm^3)	1.9	1.9	1.9
Void content (%)	1.4	2.8	----
Barcol hardness			
Avg.	60	60	65
Std Dev.	4	4	3
Barcol after DMA test			
Avg.	64	62	67
Std Dev.	3	3	3

TABLE V

PROPERTIES OF THE POLYESTER/GLASS LAMINATE

Tensile Strength (Ksi)
    ASTM D3039 (1")
        Avg.   91.3
        Std Dev.   3.4

    ASTM D3039 (0.5")
        Avg.   83.8
        Std Dev.   5.0

    ASTM D638
        Avg.   85.6
        Std Dev.   2.6

Compressive strength (Ksi)
        Avg.   32.9
        Std Dev.   0.7

Short Beam Shear Strength (Ksi)
        Avg.   5.0
        Std Dev.   0.3

Density (g/cm^3)
        Avg.   1.8
        Std. Dev.   0.0

Resin content (%)
        Avg.   31.1
        Std Dev.   0.2

Void content (%)
        Avg.   3.3
        Std Dev.   0.0

BIBLIOGRAPHY

1. M.. Coats, "Fast Online Technique Tracks Thermosets Cure", Modern Plastics, May 1986.

2. A.K. Munjal, "Manufacturing of High Quality Composite Components in Aerospace Industry", EM86-108, SME Conference, Composites in Manufacturing 5, January 1986.

3. L.I. Johnson, "Composite Standards Activities of the Society of Automotive Engineers", Journal of Composites Technology & Research, vol. 8, no. 1, spring 1986.

4. DoD/NASA Structural Composites Fabrication Guide, Section 5.

5. P.. Teagle, "The Quality Control and Non-Destructive Evaluation of Composite Aerospace Components", Composites, vol 14, no 2, april 1983.

6. J.S. Chen, A. B. Hunter, "Development of Quality Assurance Methods for Epoxy Graphite Prepreg", NASA contractor report 3531, Boeing Commercial Airplane Company, 1982.

7. T.F. Saunders, M. Ciulla, S. Wehner, J. Brown, "Application of Fourier Transform Infrared Spectroscopy for Quality Control Analysis of Epoxy Resin Prepregs Used in Helicopter Rotor Blades", SPIE Vol. 289, 1981.

8. N.C.W. Judd, W.W. Wright, "Voids and Their Effects on the Mechanical Properties of Composites - An Appraisal", SAMPE Journal, January/February 1978.

9. C.A. May, T.E. Helminiak, H.A. Newey, "Chemical Characterization Plan for Advanced Composite Prepregs"

10. J.M. Whittney and C.E. Browning, "On Short-Beam Shear Tests for Composite Materials", Experimental Mechanics, V25, N3, Sep 1985.

11. J.M. Barton, D.C.L. Greenfield, "The Use of Dynamic Mechanical Methods to Study the Effect of Absorbed Water on Temperature-dependent Properties of and Epoxy Resin-carbon Fibre Composite", British Polymer Journal, Vol. 18, No 1, 1986.

12. G. Samanni, C. Zanotti, "The Effects of Matrices Formulation on Mechanical Properties of Epoxy Glass Fiber Prepregs", Vertica, V9 N1, 1985.

Presented at the SME FINISHING WEST '88 Conference, September 1988

Statistical Control of Cyclic Corrosion Testing of Electrocoated Panels

Peter D. Clark and Lynn E. Pattison
BASF Corporation

Introduction

Corrosion testing is important to the finishing industry. Coatings are often expected to protect the underlying materials as well as serve a decorative purpose. Important decisions about which coating to select for a particular application often are based on the results of corrosion tests.

The improved corrosion performance of metal goods that have been electrocoated has greatly extended the service life of all sorts of familiar products. The current generation of cathodic epoxy electrocoatings have corrosion protective capabilities that now present difficult challenges to those who must develop tests which differentiate between such high quality products.

Corrosion testing is performed in order to predict how a particular product will perform in actual service [1-3]. The utility of any one corrosion test depends on the extent to which the test fits the following description:

   *   The test must have the same failure mechanism that occurs in the service environment that the test is designed to simulate.

*   The test must correlate with the actual service history. (In reality, this has been done in only a very few cases. Often, any "correlations" that have been made are based on only a very small number of samples that have seen actual service environments.) [4,5]

*   The test must be reproducible.

*   The test's accuracy must be understood.

*   It is desirable to have a test which can be readily run in the routine volumes required by the finishing industry today. (In our laboratories, for example, we routinely run over 2000 panels in the scab test on an ongoing basis.)

There are many corrosion tests which are used by the finishing industry today. Concern about corrosion in the automobile industry, and the warranties against both cosmetic and perforation corrosion that automobile manufacturers are now providing, have provided significant incentives to investigate corrosion testing methodologies.

The popular ASTM salt spray test is now considered to be unreliable as a predictor of how an automobile will perform in actual service. [6-10] Cyclic corrosion tests have been, and continue to be, developed in an effort to more closely mimic the corrosion mechanisms that occur in actual service. [11-17] It is generally hoped that by more closely mimicking the mechanisms that occur in actual service, that better correlation with actual service histories will result as a consequence. To date, this fact is not well established, however.

Our laboratory is actively investigating the larger issues of how well various tests mimic the "real world corrosion mechanisms" and is actively trying to make correlations between accelerated tests and "real world" service history. However, this paper will provide insights that have been gained in our laboratory about the reproducibility and the accuracy of cyclic corrosion tests. It will also provide some insight into the problems and solutions in routine testing of large quantities of panels. These insights should be important to those who are developing new test methodologies as well as those who must interpret the data from the existing test methodologies.

```
TYPICAL CYCLIC CORROSION TESTS

GM SCAB  GM TM 54-26
ASTM D2933
FORD APG
CHRYSLER CHIP CORROSION
VOLVO TEST
FORD CSCT


          FIGURE 1
```

144

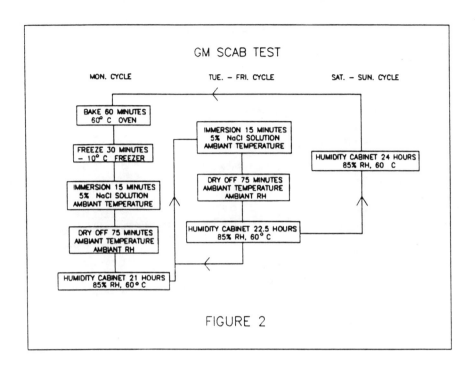

FIGURE 2

Discussion

Although there are quite a few examples of cyclic corrosion tests (Fig. 1), we chose the **SCAB** test because it is typical of many cyclic corrosion tests, and it is a widely used test. The test procedure is outlined in Fig. 2.

In order to monitor the reproducibility and the accuracy of the scab test, we have employed the techniques of **Statistical Quality Control** or **SQC**. In order to begin to understand the causes for the normal variation in the scab test process, we have begun to use the techniques of **Statistical Process Control**, or **SPC**.

We have identified a number of factors that are potential variables in the test (Fig. 3). We have attempted to control some of these factors by implementing specific test protocols, and generally paying closer attention to details.

One example of this type of factor is the transfer times. In cyclic corrosion tests, the panels are moved from one piece of equipment to another in order for the panels to be exposed to a different set of conditions. While the authors of cyclic corrosion tests typically specify how long the panels should be exposed to the various conditions within the test cycle, the time required to transfer panels between conditions, and the rates at which the new conditions are to be established are typically unspecified.

```
FACTORS WHICH ARE POTENTIAL VARIABLES
         IN THE SCAB CORROSION TEST

  *   Temperature in the humidity cabinet
  *   Relative humidity in the humidity cabinet
  *   pH of humidity in the humidity cabinet
  *   Time in each part of the cycle
  *   Purity of NaCl solution
  *   Purity of water used in humidity cycle
  *   Bacteria in either salt solution or humidity cabinet
  *   Scribe profile and depth
  *   Cool down rate in the freezer
  *   Heat up rate in the oven
  *   Temperature of the salt solution
  *   Transfer times
  *   Equipment maintenance
  *   Relative humidity of room during dry-off
  *   Air velocity and patterns during humidity cycle
  *   Incidental events

                  FIGURE 3
```

We have attempted to address this type of control problem by using equipment which allows the transfer times to be minimized. We have constructed a cart which will hold all of the panels contained within one humidity cabinet. This cart goes directly into the oven and the freezer along with the racks of panels, and thus minimizes the transfer times.

The rates at which the new specified conditions can be obtained is a function of both the capabilities of the equipment and the mass of the panels (and cart), especially when temperature changes are specified. We have instituted a regular, and routine program to maintain the equipment in hopes of trying to minimize the equipment variation. Because the demand for testing varies over time, the mass of panels has also been a variable.

Some of the variables which we have identified lend themselves to monitoring. To date, we have monitored and charted on SPC charts data on the humidity cabinet environment, and on the salt immersion solution environment.

Additional details of our test protocols have been provided in Appendix A. The actual SPC charts for the variables which we are currently monitoring have been provided in Appendix B.

The overall quality of the test results has also been monitored and charted using "control panels". In order to monitor the test, we selected 10 commercially available cathodic epoxy electrocoat products. These products were manufactured by three different paint companies. We also selected 4 different substrates: two zinc phosphated cold rolled steel (2 different suppliers of the phosphates), zinc phosphated G60 hot dipped galvanized, and unphosphated, cleaned only G60 hot dipped galvanized. These "control" panels were prepared in sufficient quantities under preparation conditions designed to minimize variations within

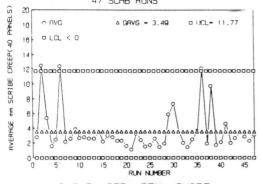

S.Q.C. AVERAGE CHART
40 PAINT/SUBSTRATE VARIATIONS/RUN
47 SCAB RUNS

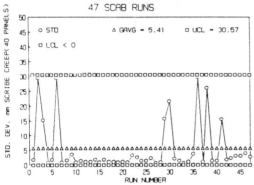

S.Q.C. STD. DEV. CHART
40 PAINT/SUBSTRATE VARIATIONS/RUN
47 SCAB RUNS

FIG. 5

This SQC chart plots the avg. and std. dev. of the 40 paint/substrate variations for each of 47 runs.

a given paint / substrate set.

As described in Appendix A, we group panels into runs which constitute the number of panels which we can load into a given humidity cabinet at any one time. Each "run" includes a set of the 40 control panels.

Results

The SQC chart shown in FIG. 5 shows how well the scab corrosion process is reproduced overall. The averages being plotted as a function of the run number are the averages of the 40 paint/substrate control panels that were run in every run. If there were no variation in the test, and if the sets of 40 panel/substrates were perfectly identical, the average for the set would be a constant, or a straight, horizontal line. Likewise, the standard deviation would never change.

As is normal for all processes, some variation does in fact occur. The SQC chart shows us what sort of variability can be expected as normal variation in the test. The grand average, is the arithmetic mean of all the averages. The upper and lower control limits define a space that is + 3 standard deviations away from the mean. If normal variation is occurring, this space would be expected to contain 99.8 % of all the results.

When points are outside the space defined by the upper and lower control limits, variation has resulted for reasons other than that expected from the normal random variability of various variables. These occurences must be investigated further. The overall scab process that has been run in our laboratories shows examples of this type of occurence.

If we look at the control charts for the different

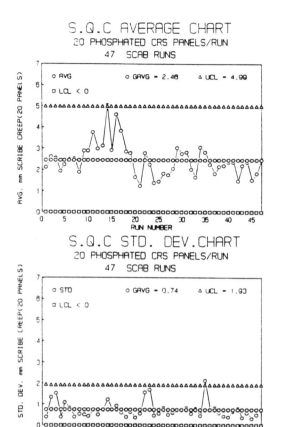

FIG. 6

This SQC chart plots the avg. and std. dev. of the 10 paints X 2 Phos. CRS substrates for each of 47 runs.

of the panels have delaminated entirely, which skews that run's average to unusually high values.

The reason that we see occurances of delamination on bare galvanized substrates is not yet clear. It is speculated that there may be two different corrosion mechanisms involved. Most of the time, the scab corrosion process might favor a mechanism which does not produce delamination on bare

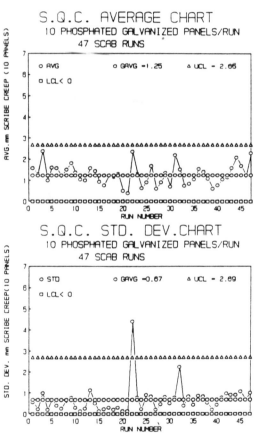

FIG. 7

This SQC chart plots the avg. and std. dev. of the 10 paints on phosphated HDG for each of 47 runs.

substrates (Figures 6 - 8), the reason for this unusual result becomes more evident. The control charts for the phosphated CRS and the phosphated galvanized appear to represent more normal processes, with normal process variation. The control chart for the bare galvanized substrate, however, represents a process that has unusual events where at least some galvanized substrates. At other times the conditions do produce delamination on some of the paints.

While further investigation will hopefully shed additional insight as to how and why this apparent difference in failure mechanisms might occur, we can conclude that this particular test is not reliable on bare galvanized substrates.

As the process is improved, and the various "variables" in the test are controlled to within tighter ranges, we can expect to see some improvements in the variation in both the means and the standard deviations.

We have taken the 40 paint/substrate variations that we have run, and

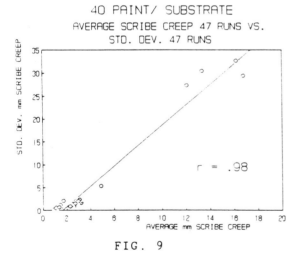

FIG. 9

The 40 std. dev. of the 47 runs of the 40 paint/substrate variations have been plotted versus the average scribe creeps for the 40 paint/substrate variations.

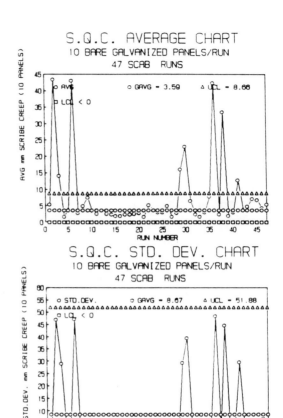

FIG. 8

This SQC chart plots the avg. and std. dev. of the 10 paints on bare HDG for each of 47 runs.

computed the average and standard deviation over the 47 scab runs. When we plot the standard deviation vs the average, Fig. 9, we see that there is a strong correlation, and the slope is greater than 1.

This data is telling us that our uncertainty in any given scribe creep result is large relative to the value of the scribe creep obtained.

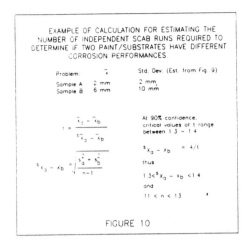

EXAMPLE OF CALCULATION FOR ESTIMATING THE
NUMBER OF INDEPENDENT SCAB RUNS REQUIRED TO
DETERMINE IF TWO PAINT/SUBSTRATES HAVE DIFFERENT
CORROSION PERFORMANCES.

FIGURE 10

This means that one must be very careful in interpreting scab corrosion results of systems which have roughly the same scribe creep value. In practice, many decisions have been made on the basis of a few mm difference in scribe creep between two panels run in a single scab test. But our data would lead one to think that this may not be a very reliable practice.

A hypothetical example for how many independent runs is required to have 90 % confidence that two paint/substrates are different is shown in Figure 10. In this example, the one paint/substrate has an average scribe creep of 2 mm and the other has an average scribe creep of 6 mm. Many people have made decisions about the corrosion of paints on differences in the scribe creep that are less than this amount, which is why these numbers were selected.

In order to have 90 % confidence that these two paint/substrates are different, assuming the variation fits the data we are showing in Fig. 9, we would require on the order of 12 independent runs! This is a fact which we believe is not appreciated by a large majority of paint scientists.

We have also looked at this data in another way. We have taken the average and standard deviation of all 40 paint/substrates within each run. When we plot these std. dev. vs. these averages, as we have in Fig. 11, we again see a strong correlation.

This is telling us that when the test conditions yield the most severe overall corrosion, the variation between samples is the greatest. If the rank order between the paint/substrates were always the same, one would

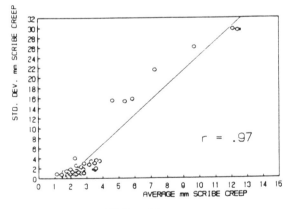

FIG. 11

The 47 std. dev. determined for the variation within the 40 paint/substrate set for each of the 47 runs have been plotted versus the average scribe creeps of the 40 paint/substrate sets for each of the 47 runs.

be inclined to suggest that the more severe tests would
differentiate better between paint/substrates. However, this
does not appear to be the case in our studies. We only
observe increased variability both overall and within each
paint/substrate group.

In our Southfield, MI laboratories, we operate 4 scab
humidity cabinets. We have made SQC charts for the scab runs
that were conducted in each of these cabinets. (Note: The
run numbers indicated in each of these graphs are the same
run numbers that were shown in Fig. 5 - 8.)

In Fig. 12 - 15, we see the SQC plots from the 4 cabinets
excluding the data from the bare galvanized panels, which, as
was seen in the previous graphs, produces less reliable
results. We observe that the cabinets are reasonably
reproducible with each other and that each cabinet is showing
the normal variation expected of a process that is well
controlled.

Inclusion of the bare galvanized data leads to the expected
increases in variability. In Fig. 16 - 19, we show the SQC
plots of the average and std. dev. of the 40 panels which
were run in each scab run.

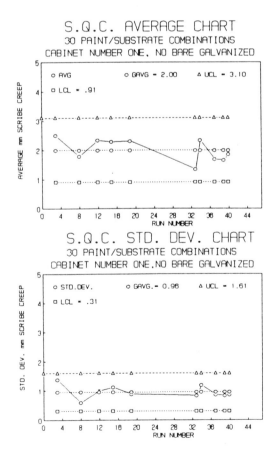

FIG. 12

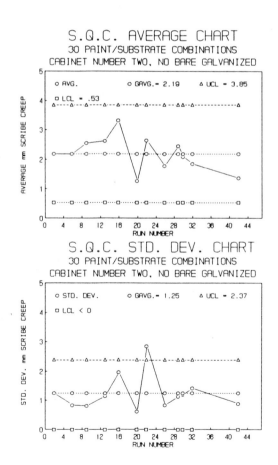

FIG. 13

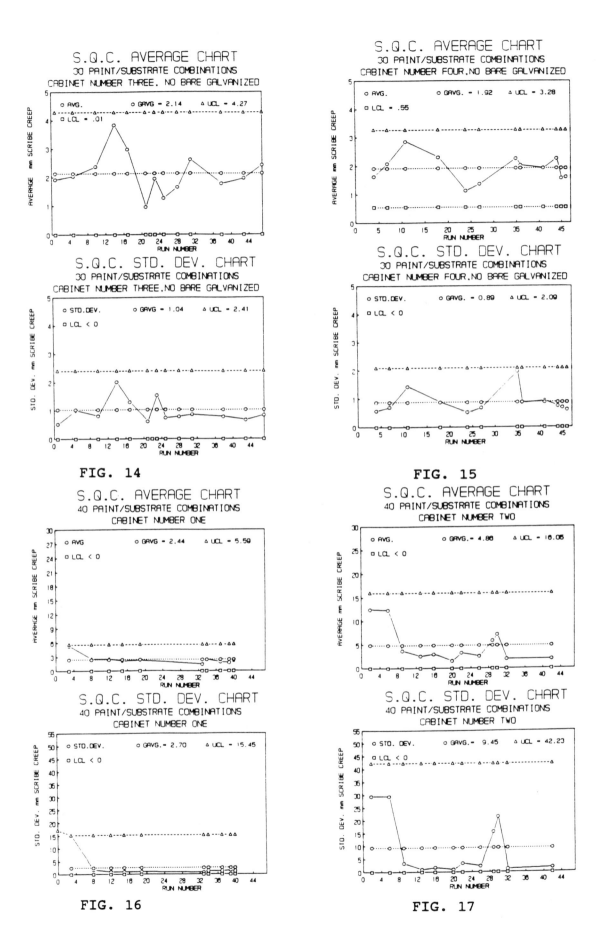

FIG. 14

FIG. 15

FIG. 16

FIG. 17

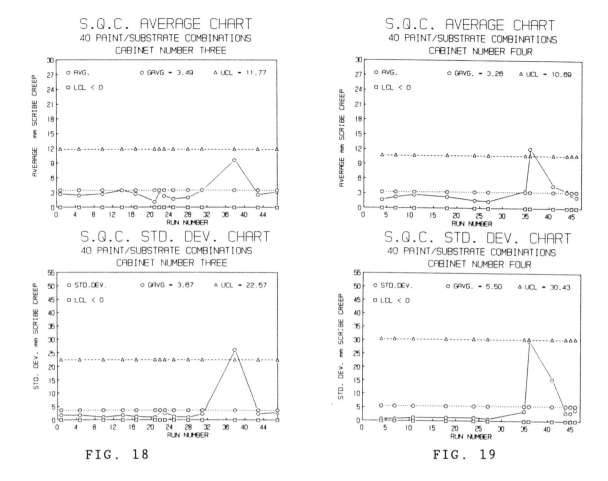

FIG. 18

FIG. 19

Conclusions

Although these results are still preliminary, we think that there are some conclusions which can be made at this time. These are as follows:

*   Statistical quality control and statistical process control of scab corrosion testing does appear to have potential for helping to make the testing process as reproducible as possible within the current capabilities of testing equipment.

*   The scab test itself appears to be unreliable when cathodic electrocoatings are coated on bare galvanized substrates.

*   The scab corrosion test does not appear to be capable of differentiating between different cathodic epoxy electrocoats on phosphated substrates unless large numbers of independent runs are made.

*   The test does differentiate between the different

HDG. We do not have data, however, that would indicate that we could differentiate between different phosphates within a general class of phosphates, ie. two different zinc phosphates.

*   New cyclic corrosion test developers need to understand the consequences of having the test adopted for routine testing, should provide some measurements of how reproducible the method actually is, and should provide some insight into how well the test can differentiate between similar coating products. Ideally, correlation of significantly large amounts of panel tests should be made with duplicate panels exposed to the "Real World" conditions the test is designed to simulate.

APPENDIX A
SCAB CORROSION TEST PROTOCOL

Panel Protocol

* All panels were 4" x 12" panels. Panels had a punched hole for hanging the panels on racks. The holes were punched prior to phosphating and electrocoating.

* The panels were given a 10 inch center scribe before being placed into cabinets.

* The panels were scribed by hand using the scribing tool specified in GM 9102 - P. (Note: Development of techniques and equipment to machine scribe the panels with very precise depth control is under way in our laboratories.)

* Panels were hung onto a fiberglass rack which was designed to allow the panels to hang vertically at all times. Panel to panel contact is carefully avoided.

* Panels are placed into test without taping of edges.

Humidity Cabinet Protocol

* The humidity cabinets used are Lunaire Model GEO632M-3 or GEO932-3M. These cabinets have a stainless steel interior, and a side to side air flow pattern.

* Panels are placed into the cabinet such that the racks are aligned with the air flow stream, and the panels are thus perpendicular to the air flow stream.

* The humidity in the cabinet is regulated by a wet bulb / dry bulb humidity controller. The actual humidity in the cabinet is monitored using a Vaisala electronic humidity meter on a weekly basis.

* The socks on the wet bulbs of the cabinet humidity controller is replaced every 2 weeks. The cabinet is rinsed out with deionized water on a weekly

basis.

* The temperature of the cabinet is monitored with an independent, calibrated thermometer on a weekly basis.

Salt Solution Immersion Protocol

* The salt solution is replaced on a weekly basis. All rust and debris (resulting from the corrosion of the panels) is removed.

* The salt solution is monitored for pH, specific gravity, and temperature daily.

Cycle Time Protocol

* All transfer times between cycle environments are accomplished in 2 minutes or less.

* All cycle times are $\pm$ 1 minute.

* Times in oven and freezer are resident times, and not times at the specified temperature. Thus the heat-up or cool-down time is included in this time.

* All transfer times, or unavoidable delays, are taken from the time in the humidity cabinet. Since panels spend the greatest percentage of time in the humidity cabinet, any variation in time will have the least impact on the overall test if the variation is restricted to the time in the humidity cabinet.

Run Protocol

* Panels are grouped into a "**Run**" of panels which is started on the first Monday of the test beginning with the oven cycle. The run group will have no more than the maximum number of panels a single humidity cabinet can hold without panel to panel contact.

* A run of panels is tested all at the same time through every phase of the cycles. This means that all panels are placed into a single humidity cabinet, all panels go into the same salt immersion solution at the same time, and all panels go into the freezer and the oven together at the same time. To the extent it is possible, the panels experience the same, identical environment throughout the test.

* Control panels are placed into test along with the

APPENDIX B

SPC CHARTS FOR MONITORED VARIABLES IN SCAB CORROSION

Humidity in the cabinets:

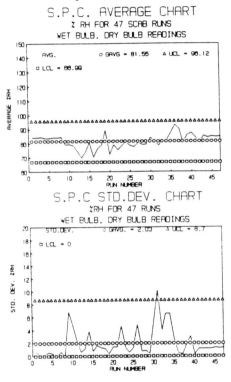

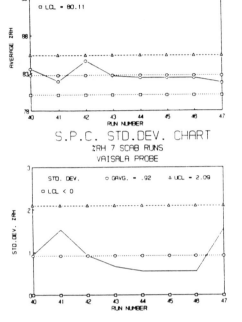

Temperature in the cabinets:

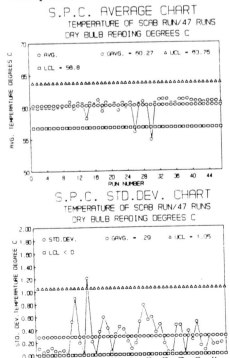

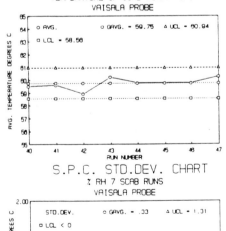

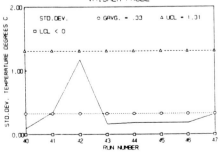

157

experimental panels. These control panels must
have sufficient history in test to be able to
indicate if the test itself has had a problem.

Temperature Profile

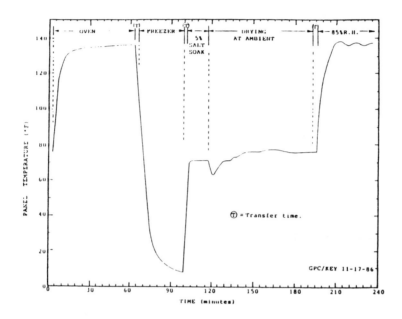

Thermal history of a typical panel set during
the Monday cycle.

Total number of panels in the run:

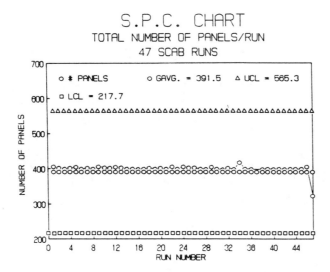

REFERENCES

1. Corbett, R. A., **Chemical Processing**, Oct. 1987, P. 28 - 32.

2. Liedheiser, H., **Corrosion**, 38, 7, P. 374 - 383.

3. Funke, Liedheiser, Dickie, Dinger, Fischer, Haagen, Herrmann, Mosle, Oechsner, Ruf, Scantlebury, Svoboda, and Sykes, **Jour. Coat. Tech.**, 58, 741, 1986, P. 79 - 86.

4. Blekkenhorst, F., Nagel Soepenberg, E., Roelofsen, M., Schoen, J. P., **Corrosion 88**, March 21 - 25, 1988, Paper No. 354, NACE.

5. Vrable, J. B., **SAE Paper 850002**, Feb. 25 - Mar. 1, 1985.

6. Funke, W., **Corrosion Control By Organic Coatings**, H. Liedhieser Ed., Science Press, 1979, P. 35-45.

7. Liu, T., **Corrosion Control By Organic Coatings**, H. Liedhieser Ed., NACE, 1981, P. 247-254.

8. Floyd, F. L., Groseclose, R. G., and Frey, C. M., **JOCCA**, 11, 1983, P. 329 - 340.

9. Lambert, M. R., Townsend, H. E., Hart, R. G., Frydrych, D. J., **Ind. Eng. Chem. Prod. Res. Dev.**, 24, 3, 1985, P. 378 - 384.

10. Standish, J. V., **Ind. Eng. Chem. Prod. Res. Dev.**, 24, 3, 1985, P. 357 - 361.

11. Jones, D., **ACS Polym. Matls - Sci. and Eng.**, 53, Fall 1985, P. 470 - 474.

12. Wyvill, R. D., **Metal Finishing**, Jan. 1982, P. 21 - 26.

13. Knaster, M., and Steinbrecher, L., **ACS Polym. Matls. - Sci. - Eng.**, 53, Fall 1985, P. 703 - 708.

14. Hospadaruk, V., Huff, J., Zurilla, R. W., Greenwood, H. T., **SAE Paper 780186**, 1978.

15. Stephens, M. L., Davidson, D. D., Soreide, L. E., Schaffer, R. J., **SAE Paper 862013**, 1986.

16. Smith, A. G., **ACS Polym. Matls. - Sci. and Eng.**, 58, June 1988, P. 417 - 420.

17.   SAE .J1563, May 1987.

BIBLIOGRAPHY

1.   Runyon, R. P., and Haber, A., **Fundamentals of Behavioral Statistics**, Addison-Wesley, 1971.

2.   Hunt, P. J., **Statistical Process Control**, Productivity Management Consultants, 1987.

ACKNOWLEDGEMENT

We would like to thank Dr. K. Yee and Mr. G. Callewaert for providing the temperature profile of the scab test which appears in Appendix A.

Presented at the SME FINISHING '87 Conference, September 1987

Computer-Aided Quality Control Gives Quality and Productivity Gains for Metal Finishers

A.J. Bates
DATAPAQ, Limited

For the finishing industry, the challenge of quality and cost is particularly relevant. Time and time again it is the Fit and Finish of a product as much as its technical performance that is shown to have a major impact on customers' purchasing decisions.

Given the title of this session, "Statistical Process and Quality Control", this challenge is very relevant. Control relates directly to this ultimate paint and finish quality, and cost, of the finished product. In order to properly control a process we implicitly assume several factors are known:

o   Precise definition of the process to be controlled.
o   Choice of unit(s) to measure that process which can be related to process parameters such as heat, time, pressure, etc.
o   Ability to easily acquire and interpret those measurements, and
o   Act, based on the information gathered, to keep the process in control rather than to "react" when the process is out of control.

Over the past decade, robotics, control systems, material handling, rapid color changes and oven manufacturing have all improved dramatically the finisher's ability to control the process. And improvements are continually being made. However, one very large part of the coating process, and typically one of the larger bottlenecks of the process, is the ability to monitor and control events occurring within the curing oven.

Because of the hostility of the oven environment, lack of proper coating specifications and the inherent "mystery" of events within the oven, the monitoring and control of the oven curing process has made little, if any, progress until recently.

Large amounts of time, money and capital investment are used to get a workpiece ready to enter an expensive oven in order to cure the coating. It is often time the LAST manufacturing process, aside from shipment, which that workpiece undergoes. A mistake at this point is indeed costly. However, only AFTER the finish of a product goes out of the range of acceptability is attention paid to the source of this deviation. Various methods, based on the physical appearance and ruggedness of the finish, and all after the fact, have been used to measure the cure in a production environment. The possibility of avoiding this unacceptable cure is discussed continually. However there has been no way to quickly and easily monitor the degree of cure of workpieces right out of the oven based on the properties of the coating. Not only does this result in unacceptable product, high energy costs, and lost production time, but often hard to resolve disputes occur among those involved--the paint manufacturer, the oven people, process control, manufacturing and quality control.

Given these problems of definition, measurement and expensive mistakes, it is necessary to define the process we would like to evaluate and control, decide upon a unit of measurement, and see if it is possible to monitor and control that process in a cost effective manner. While most firms have the technology and capability to collect a great deal of data it is only by processing the relevant data into simple discrete packages of information that systems can be developed to ensure that the right balance between maximum output and maximum quality is attained at a competitive cost.

BACKGROUND

The Finishing Industry has been striving for a "perfect" cure since the advent of thermoset polymer based coatings.

As suggested in the name thermoset, there is a fixed end point at which all the possible cross linking has taken place. The system is set, and any further application of heat will not have any beneficial effect on the system.

With a thermosetting paint system, be it acrylic, polyester, epoxy or alkyd, the cross linking process can take place at relatively low temperatures at an infinitesimally slow pace. The underlying principle of the curing process is one of catalyst, or "curing agent", e.g. TGIC for polyester powders or amine derivatives for epoxy resins.

Paint can be formulated with varying proportions of polymer and catalyst to achieve almost any desired curing time at a given temperature. However, there are many other criteria that have to be satisfied during the curing process that limit the actual range of curing times within each general polymer type. For example, the degree of flow and gloss achieved may be directly related to the rate at which the curing process proceeds. Similarly there is a maximum temperature over which the paint film may be degraded or the pigments discoloured. In most systems there is also a minimum temperature below which no effective curing can take place or where the physical properties acquired in the final paint film will not be properly developed.

These upper and lower limits define the boundaries of normal operation for the paint curing process. <u>HOWEVER, THERE IS ALWAYS A CONSIDERABLE RANGE OF TIMES AND TEMPERATURES BETWEEN THESE LIMITS AT WHICH CURING CAN, AND DOES, TAKE PLACE.</u>

The rate of reaction of the curing mechanism varies with temperature, but typically this relationship is not linear, and usually approximates a First Order Reaction e.g., a rise of 10 C usually doubles the cure rate. The following three curves illustrate three different theoretical "temperature profiles" which would cause an equal, and full, 100% cure of a thermoset paint.

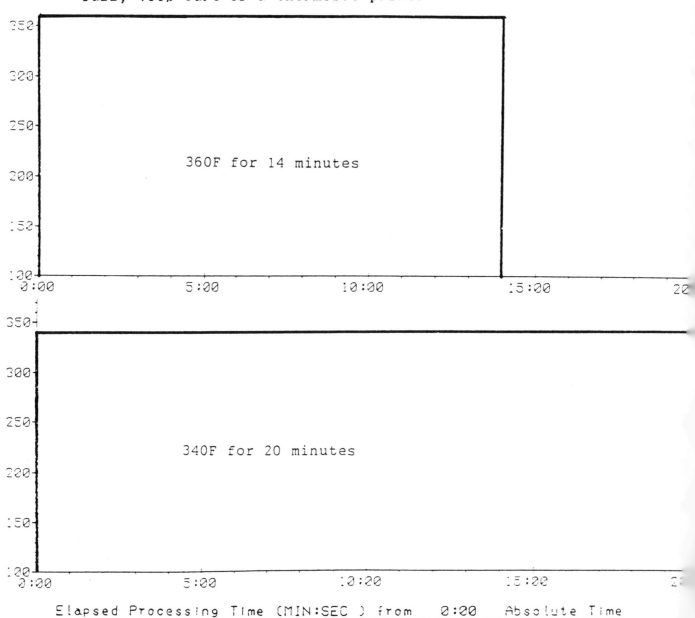

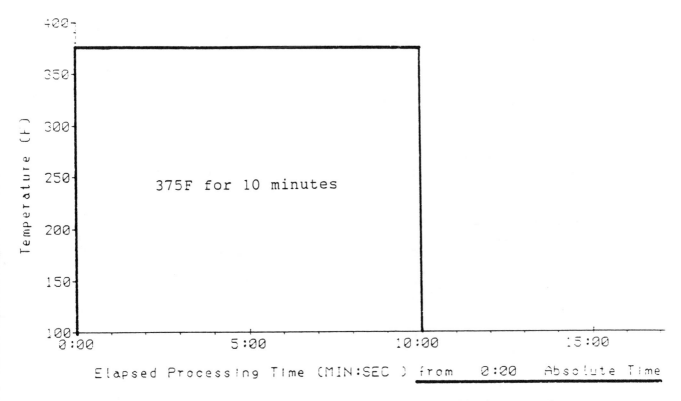

One can see graphically that usually small temperature changes make a large difference in cure time. In an industrial world, with workpieces of varying thicknesses, oven balance, etc., no one could expect such a "rectangular" temperature profile on a workpiece they are trying to cure. However, given these profiles we can deduce the following cure curve:

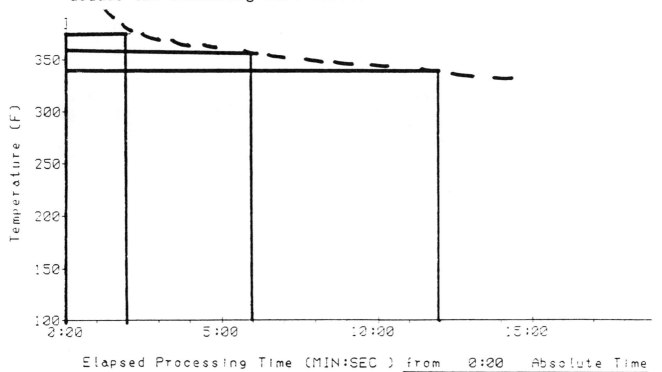

Any combination of time and temperature found on that curve would yield a 100% cure. Let us now return to our problem - monitoring and controlling the process of cure.

WHAT IS THE OBJECTIVE OF THE PROCESS WE WANT TO MEASURE?

We want to measure how well the oven is performing its objective - the heating of the substrate in order to cure the coating.

The oven facility is one of the most capital, space, and energy intensive pieces of equipment in many manufacturing plants. Zones are set, air temperatures measured, efficiencies tested, etc., but often the basic purpose of the oven is forgotten. The OVEN'S PURPOSE is to heat the finish so that adequate flow and crosslinking take place. Most ovens heat the finish through first heating the substrate. The various shapes, sizes, and densities of workpieces, as well as oven variations, cause wide differences in this transfer of heat process. The next step is to measure to what extent the transfer, and thus, cure occurs.

The question becomes, how do we measure if the finish is completely cured regardless of the substrate? And we also want to be sure that certain maximum temperatures have not been exceeded.

WHAT IS THE PROPER UNIT OF MEASURING THIS PROCESS?

We are measuring the cure of the coating. Thus, a simple Index of Cure which can combine both heat and time into a single discrete number would be ideal. With such an Index we could compare more readily the cure on a 1/2" piece of steel to the cure of a 1/8" piece of aluminum based on their individual temperature profiles.

The rationale for this becomes more clear when parts with varying degrees of thicknesses or heat transfer capacity pass through an oven in different locations. The objective is simple to define - make sure the coating on the thickest section is cured adequately while minimizing the over cure on the thinner sections. An Index of Cure allows this comparison to be made. If a "perfect" cure is summarily indexed at 100, we know the thick section must have an "Index of Cure" of equal to or greater than 100. If it goes substantially over 100, less heat/time would be necessary. Less heat would reduce energy (or one could speed up the line) which minimizes the overcure on the thinner substrates passing through the oven at the same time.

The first two decisions of monitoring the effectiveness of the oven are now known. We have defined the objective of the oven (cure) and the unit of measurement (Index of Cure). However, this is contrary to most methods of evaluating ovens and cure. Historically, time at temperatures ("I need 15 minutes at 390 and I don't care about anything else") has been the basis of evaluating if the oven was achieving its objective. There are three reasons for this - It's understandable by both paint shop and coating suppliers, it was the only easily obtainable method of evaluation and "that's the way we've always done it". Of the three, the third is quite dangerous, as many U.S./British, etc. industries can testify. The first two are quite real, and must be responded to. Every so often an engineer with 4-8 hours on his hands may decide to evaluate the cure on different substrates. He would painstakingly transfer times and temperatures from different profiles, assess cure at each point (if the data were available from the paint supplier) on each curve, run the totals and, by the time he/she was finished, not want to do it for another year or so! This leads us to the next step.

ABILITY TO EASILY ACQUIRE AND INTERPRET THE APPROPRIATE PROCESS PARAMETERS.

Currently most coaters use, if anything, smoke charts or mechanical recorders placed in fragile vacuum flasks. The resulting graphs, which 25 years ago were a huge advancement over heat tapes, have now been found to be suspect in temperature accuracy, painful to analyze, fading rapidly through age, and almost impossible to accurately analyze, given that a 10 C change in temperature doubles the cure rate. This 10 C change can almost be wiped out by a thick pencil line on some mechanical recorders, that are only accurate to +/- 6 deg.C. anyway!

With the enormous advance in computer hardware and software, one can gather temperature data, and store it digitally for later analysis by software. The software can even make the data more accurate through linearization. With these electronic developments, plus development of new insulations, it is possible to send computer based temperature gathering equipment weighing less than 7 pounds in total through a 400 F oven for an hour with no ill effects.

Of course, the gathering of temperature data, whether easily, accurately or safely, has been done over the past three decades with regularity. However, in this age of Statistical Process Control and demand for efficiency combined with high quality, coaters can no longer rely ONLY on the paint ombudsman who will look at a graph, rub his chin, shake some rattles and bones and issue a pronouncement of what is to be done to the oven. Experience has no substitute, but if we can gather temperature data and store it digitally, then computer software programs can analyze that data in the manner we wish, based on the properties of the coating and characteristics of the oven. Ideally, this analysis should be on a regular basis. What should this analysis include?

o A multi color graph so that varying substrates may be visually differentiated and easily studied.

o Curing specifications of the coating.

o For each probe, an "Index of Cure" so that there can be a quick spot check to see that each substrate is cured and energy is not being wasted by over curing.

o Warning signals if the oven or a substrate are becoming too hot.

o A quick notation on the peak temperatures reached.

o "Time at temperatures" for each probe. While the importance of this is decreased by the Cure Index, it is still an important parameter.

o Time of run, oven settings, operator, coating used.

With the use of computers, the above information can be printed and stored within minutes after the data has been gathered during an oven run. And through programming based upon use by production personnel, this analysis can be completely carried out by a production person. With time, acceptable process parameters are decided upon. When those boundaries are crossed (or approached), a more senior person would become involved. Thus while the computer is analyzing the data (Index of Cure, time at temperature, etc), it is still up to those in the paint shop to interpret what the analysis means. The key step is that the temperature profiles are analyzed, and analyzed accurately, and the interpretation is based on that analysis as well as the experiential analysis of looking at the curve. The computer aided analysis of the temperature profile takes much of the mystery out of the curve.

ACT TO KEEP THE PROCESS IN CONTROL

While temperature recorders/analyzers have been used mostly
as cursory oven checks or to help solve a problem which has
already occurred - and cost money - , the major benefit, if
used correctly, is to monitor the process and avoid costly
over or under curing conditions.

Let us take two examples. In the first, a single piece of
multiple thicknesses is being cured. In the second, an
aluminum extruder is changing from a thick to a thinner
extrusion. In example one the piece passing through an
oven:

    o is of varying thickness

    o has upper and lower control limits, set for their
      index of cure on each of two critical points.

Point	Cure Index Lower Limit	Cure Index Upper Limit
1. Thick point	105	120
2. Thin end	115	140

A chart is arranged on the wall as follows:

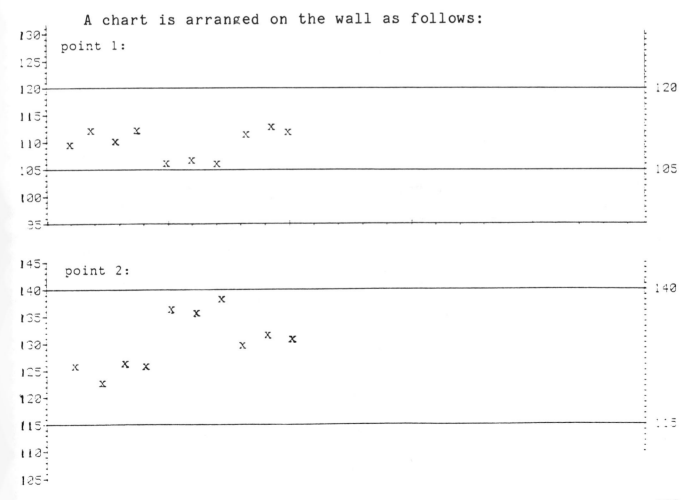

While both are within acceptable boundaries, simple visual inspection can see that each Index of Cure hovered dangerously close to acceptable limits. The oven has been set up, balanced, and tested such that these control limits, if the oven were working properly, should not be reached. Upon examination, it was found that a mechanical malfunction had caused air to be redirected, resulting in the change. It was promptly fixed and the Index of Cure settled more comfortably in the middle range. If only the normal "Time at Temperature" had been used, the problem would not have been seen, as the times remained the same. It was a case of the higher temperature, where curing occurs much more rapidly, that the difference could be seen.

Example 2 shows how, by using the Index of Cure, a coater can rapidly approach a proper cure while minimizing energy (or maximizing oven capacity).

The spec for this powder was 390 for 10 minutes (or 375 for 12 minutes, or 410 for 7 minutes). The extrusions, regardless of thickness, passing horizontally through the oven, had the following control limits:

Extrusion	Cure Index Lower Limit	Cure Index Upper Limit
Top Position	120	140
Middle Position	115	135
Bottom Position	110	130

While the "historical" spec for 10 minutes was never reached, the Cure Index would constantly fall between these limits, set by the process engineer as that with which he would feel comfortable with. When a run of thinner extrusions would enter, the temperatures would be decreased so that the Index of Cure would still remain within the proper boundaries. When thicker extrusions were about to enter, the opposite would occur.

If the line speed were increased, the oven settings would be also - with the goal being to remain within the upper and lower limits of the Cure Index. Below is a sample of the chart tracking the TOP positions.

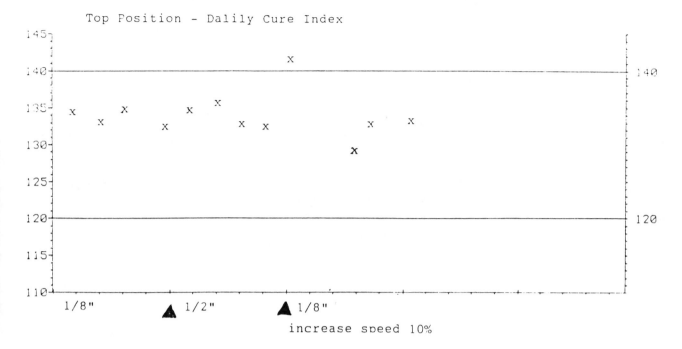

Top Position - Dalily Cure Index

Notice how, regardless of line speed or extrusion
thickness, the control limits were respected. Also, as the
line speed increased, oven temperatures were increased.
The Upper Control limit was passed, but quickly brought
into line.

CONCLUSION:

Through the use of up to date microprocessor technology and
computer software development, it is now possible to
accurately and easily assess the cure of workpieces,
regardless of substrate characteristics. This assessment
allows the coater to:

    o insure adequate cure
    o minimize overcuring and undercuring
    o reduce set up time
    o minimize energy costs
    o reduce warranty claim
    o balance oven
    o maximize oven capacity by understanding precisely the
      cure needed and achieved.
    o provide precise historical records

It is time for the cure oven to become a part of the
monitorable production process. It can be done easily and
accurately. To assess the cure based on the properties of
the coating is the most logical way in which to monitor and
control the cure oven process.

Presented at the SME/FMA FABTECH INTERNATIONAL Conference,
September 1987

Statistical Process Control for Tube Fabrication—An Update

David A. Marker and David R. Morganstein
Westat Incorporated

This paper discusses the importance of upper management's role in the implementation of statistical process control. Why are so many companies concerned with statistical thinking? What can it do for my company?

We first describe what we mean by statistical thinking, and what is involved in measuring the cost of quality. After comparing the traditional view of quality with the newer "Japanese" view, we cover basic statistical definitions, elements of statistical training, and the need for standards.

Many companies have turned to statistical techniques in an attempt to avoid going out of business. Others have become involved in response to pressure from customers such as Ford and GM. Still others have noted that while the West has not stood still in respect to quality, over the last 30 years the reputation of Japanese goods has gone from "use it today before it breaks" to "the very best quality for the dollar." What has happened? What do they know that we don't?

The effects of this quality differential can be seen quite clearly in the following table comparing the inventory costs of Japanese and American automobile makers. The Japanese typically have a 10-part per million (ppm) on-line rejection rate, compared to the 1 to 4 percent of American car companies. Both make cars with approximately 10,000 parts. The low Japanese defective rate results in an extra billion dollars of inventory needed to produce American car!

Table 1. Comparison of Japanese and American car manufacturing

Japan	U.S.
10-ppm reject rate on-line or 0.001 percent defective	1 to 4 percent defective
10,000 parts per car	10,000 parts per car
1 part per 10 cars replaced on-line	100 to 400 parts per car replaced on-line
1 to 6 hour spare parts inventory	1 to 3 month spare parts inventory
$0 in inventory	$1 Billion in inventory

In response to these differentials, many Western companies have started to examine the Japanese methods and have begun to realize that they too must incorporate statistical thinking into their decision-making if they are to compete in the international market place. Both Ford and GM now require of their suppliers an on-going plan demonstrating management's commitment to quality, including the presentation of control charts accompanying each order. This in turn has led the management of all their suppliers to try to discover the secrets of statistical process control.

The first book to discuss in detail how to implement Statistical Process Control (SPC) was Walter Shewhart's 1931 book, Economic Control of Quality of Manufactured Product. The role of management in SPC is described in W. Edward Deming's Quality, Productivity and Competitive Position. Many of the ideas included in this paper are also discussed in these books.

John Johnston, the former Director of Research at U.S. Steel, said:

"The possibility of improving the economy of steel to the customer is therefore largely a matter of improving its uniformity of quality, of fitting steel better for each of the multifarious uses, rather than of any direct lessening of its cost of production."

This would not be particularly noteworthy except that Mr. Johnston made this comment during the 1930's! The ideas of SPC are not new; it is only the growing recognition of their applicability that is new.

COST OF QUALITY

Many companies have begun to compute their cost of poor quality. There are many visible costs such as downgrades, rework, and scrap, but there are also a great many less visible costs such as inspectors and downtime. Still, other costs are impossible to accurately measure. For example, how much did you lose last year due to business lost or never acquired because of a bad reputation? Feigenbaum, in a recent issue of Quality Progress estimated that while a happy customer tells eight others, an unhappy one tells 22 others!

FIGURE 1.

Many companies in the West have estimated their cost of quality in the 10 to 15 percent range, with some up near 20 percent. (The Japanese typically consume only 2 to 4 percent, most of which covers training and data collection.) Thus, a company with gross sales of $1 billion had a cost of quality of between $100 and $200 million. What could you do with that money? Are there any capital investments that you think you need that might be funded through quality savings? Do you have an analysis that clearly demonstrates whether your current equipment is being operated most efficiently?

It should be clear to anyone that the potential savings far exceed any increased costs that may arise in the implementation of statistical process control. Figure 1 demonstrates this point graphically. As you put more of your efforts into a SPC

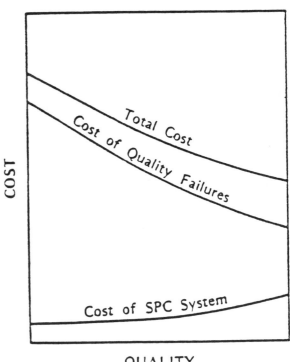

program (moving to the right in the diagram), the increased quality will result in greater and greater net savings.

WHAT IS QUALITY?

We have discussed the concept of quality, and we have talked of measuring its cost, but we have not defined what we mean by quality. Traditionally, quality has been viewed as conformance to standards. Zero defects.

Figure 2 shows a sequence of five diagrams. The first demonstrates the traditional view that any product within the specifications is fine, and any beyond the specs is totally useless.

In reality, our loss is more likely to follow the function shown in the second picture; there is no loss if we hit the aim perfectly. There is an increasing loss as we deviate further from the aim.

What does this imply for our definition of quality? Everyone would agree that the third diagram of Figure 2 indicates poor quality. But the above loss function implies that while the fourth diagram shows good quality (zero defects), the fifth shows even better quality! We will continue to improve the quality only by continually reducing the variability. This is the reason for Ford Motor Company's emphasis on **Continuing Process Improvement**, not just process control.

RESPONSIBILITY FOR QUALITY: OLD AND NEW PERSPECTIVES

We have already mentioned the need to modernize the equipment in our plants. This requires large amounts of capital. How do we know whether the gains to be anticipated through improved technology will be worth the expense? Which investments will have the higher return? In terms of quality, which will cause the greatest reduction in variability?

To answer such questions it is necessary to know what the present production system is capable of producing.

FIGURE 2.

SPECIFICATIONS

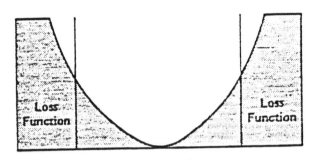

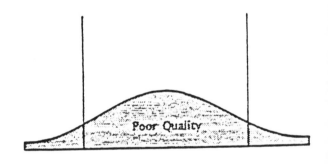

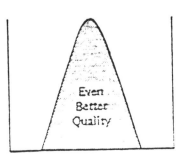

It is also necessary to be able to measure the effect of a change in the one process that is to be replaced. Is it possible to measure such changes in your company? To do so, you must be taking measurements at every major step in the production process. Only then can you measure the effect of a change to one process. It is for keeping track of just such measure-ments that a control chart is designed. The chart consists of plots of average values and ranges from small samples taken over a period of time. Using the chart, it is possible to understand how much variability is built into the system, and how much of it is a result of special, unique causes.

Recognizing that much of the variability is built into the processes of production, and therefore should not be blamed on the hourly worker (indeed Dr. Deming estimates that over 85 percent of all variation is a responsibility of management), implies that quality is everyone's responsibility, from the chief executive officer all the way to the hourly worker. Thus, statistical tools must be taught to everyone in the company if variability is to be measured and reduced.

Where does variability in the production process come from? There are many possible sources. There may be multiple streams in a casting, heating or rolling operation that are not producing equivalent products. Are the temperatures and pressures the same in each stream? Similar micrometers often produce different measurements; the same micrometer can even be read differently by two employees. Can the micrometers presently in use measure to the accuracy required by the customers?

Even if there is only one machine, there are still many possible sources of variation. Is preventive maintenance performed to ensure reliability of the operation? Do the operators on each turn run the equipment in the same way? Since they usually have no way of knowing whether the process is behaving properly at the beginning of their shift, they typically begin the day by "tweaking" the process until it meets their preferred operating conditions. If the system was already under control, all they have done by tweaking is to increase the variability of the downstream product. The result of all this variability shows up as different billet weights, tube diameters, eccentricities, etc.

Figure 3 demonstrates the traditional view of a production process, where station 1 might be a caster, station 2 extrusion presses, and stations 3 and 4 finishing processes. Alternatively, these could refer to four successive procedures in the casting process. The important point is that traditionally a product is made and then it is inspected for quality. What happens when the inspected product is rejected? Everyone is demoralized and tries to adjust to prevent it from happening again. How does anyone know which process needs to be changed and which should be left alone? Without data on each individual process, you cannot determine the appropriate response to prevent a recurrence of the problem. Similarly, without adequate data it is impossible to decide where to make major capital investments. The end result of this traditional view of quality is the continued production of scrapped tubes, coils, bars, etc.

FIGURE 3.

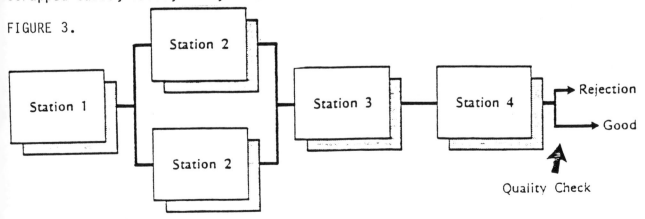

Another result of traditional quality control is a large inspection department. How much product did your inspection department make last month? How many inspectors does your company have? How many of your customers have to inspect your product? Who pays for these inspectors?

Many companies have large maintenance departments to handle all of the unexpected repairs. How much does this downtime cost your company? Typical maintenance strategy is to fix it when it breaks. Use of statistical procedures to monitor equipment will allow you to develop a system of preventive maintenance which will drastically reduce downtime.

We have all seen product with hold stickers on it resting below a sign reminding workers to "make it right the first time." If any operator has no better tool than the traditional view described above, then we are resigned to forever seeing this unfulfilled promise. Management must supply the work force with a better tool with which to do its job.

Quality cannot be left to a Quality Control Department. Instead, everyone must view quality as their own responsibility. This modern view is demonstrated in Figure 4. We have the same four stations as before, only now data is collected at each station, not just after the product is finished. The little curves by each station represent the measured variability at that step in the process.

FIGURE 4.

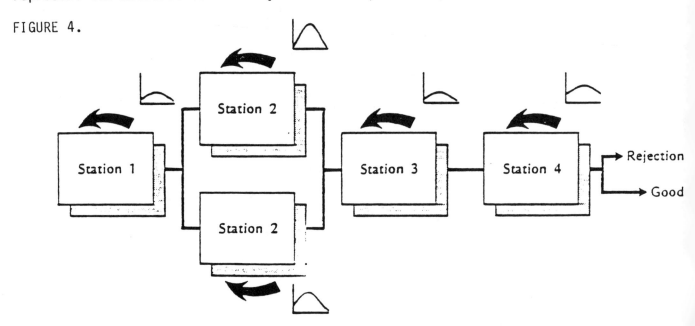

At each station people must monitor the processes that they have some control over, not the incoming materials or outgoing product. At an extrusion press, you could monitor the temperature of the billets, the quality of the lubricants, and the condition of the dies. You do not want to wait for a curved tube to be produced and then try to straighten it out. Dr. Deming compares the logic of producing unacceptable product and then trying to rework it with, "You burn the toast and I'll scrape it!" It clearly costs less to produce a straight tube than to fix a curved one.

EXAMPLE

A tube producer for an automobile company was using inspection to ensure the delivery of a product that was within the specifications. The customer knew enough statistics to recognize from the truncated distribution of diameters on received products that approximately 20 percent of all tubes being produced by the supplier were beyond the upper spec. The customer also knew enough to realize that somewhere down the line he was being charged for this waste and the 100% inspection being used to try to weed out the unacceptable product. He demanded that the producer improve his production process to eliminate the 20 percent. The supplier first realized that the hand micrometers they were using were too inconsistent to provide the necessary accuracy; they replaced them with digital mics. Then, they brought the system into statistical control by using statistical control charts on the production line. With the aid of additional statistical techniques, they were able to identify major sources of variability, eliminate them, and reduce the reject rate from 16 percent down to 4 percent. This has resulted in annual savings to the tube producer in six figures. The costs of the statistical analysis and digital micrometers were miniscule by comparison.

STATISTICAL TOOLS AND TRAINING

Figure 5 shows the monthly production of three operators of equal quality, each of whom averages 20 units per hour. Do you consider these operators of equal performance? Very few would like to have Joe working for them. Some prefer Frank, but most want to work with Charlene. Why? What is it beyond the average product that concerns people? Joe is so unstable that you have no confidence in his production for next month. Frank is extremely stable and predictable. Charlene's consistent improvement allows us to predict better future production levels for her than the others. Variability and trends over time, along with our ability to understand their causes, are at least as important in our choice of operators as is their average production. We will only be able to understand these characteristics after learning how to use and interpret a few statistical tools.

FIGURE 5.

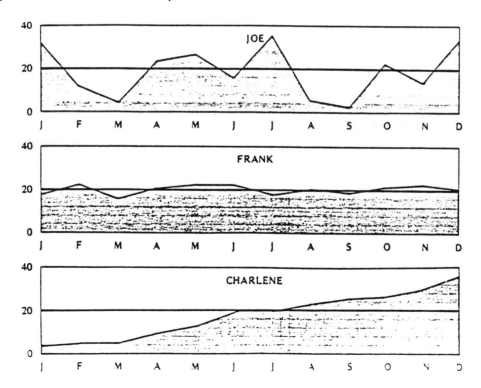

Similar confusion often arises when we try to interpret data on rejection rates. Rejections for a particular company had been averaging 8 percent and now average only 5 percent. After a period of celebration, it is noticed that rejections have suddenly reached 9 percent. How often have you heard someone say, "I thought we solved that problem six months ago!" What went wrong? Was 5 percent really statistically different from 8 percent? By this we mean that if traditionally we varied between 7 and 9 percent, then a period of only 5 percent rejects may represent a true improvement. If, traditionally, rejects were between 4 and 12 percent, then the 5 percent rate did not indicate a change in the system of production. A change in an average is only significant (worth celebrating) when it is large compared to the variability present in a stable system. How much time does your company spend responding to morning calls demanding an explanation for the rise in rejects from 8.2 to 8.4 percent? Without using statistical techniques to track the variability in the process, you cannot tell whether this rise has a specific assignable cause that you could identify, or whether it is simply due to the common cause variability of the system.

The purpose of statistical training is to develop an objective procedure whereby each employee can understand the data produced by the processes he or she is responsible for, and can interpret them in a manner agreeable to all in the company. Training, however, is only a first step; procedures must be developed to implement the training into the day-to-day corporate operations. Only in this way can full advantage be made of the statistical techniques.

The level of training should vary with the responsibilities of each employee. All levels of management must take part in sessions designed to familiarize them with the basic statistical techniques such as measuring variability and using control charts. This is important for two reasons: management must begin to use these tools as part of its basic decision-making process, and it is an excellent way to demonstrate the level of management's commitment to SPC to other employees.

Other salaried employees need to undergo a more thorough training in statistical techniques. They must understand a variety of different types of control charts, as well as how to set up and interpret them. Some of these will become internal consultants and teachers, and they, therefore, must be aware of more advanced statistical techniques such as regression, design of experiments, and reliability theory.

Hourly employees should at a minimum be exposed to the ideas of a control chart, how to plot and interpret points. Many companies will find these separations unnecessary and will train hourly employees similarly to salaried ones.

Fundamental statistical concepts should be taught to all employees. Most people do not recognize that they generally are working with samples which differ from the actual process. They do not know by how much a sample can differ from a population nor do they know how sampling error decreases with sample size. Many people continue to think that they should "count everything" even when this is physically impossible, for example, when measuring temperatures or physical dimensions.

The use of histograms and other graphical techniques to summarize large data sets needs to be encouraged. No one examines all of the numbers on a computer output; a picture does a far better job of conveying the important points. Save computer outputs for detailed searches for the cause of an irregularity.

The main shortcoming of histograms is that it is hard to compare two of them in a few words. Therefore, we must employ summary statistics that allow us to compare the histograms from two or more samples. We use the mean or median to estimate the center of thee distribution, and the range of standard deviation to estimate its spread.

CONCLUSION

Quality can only be improved through the use of statistical techniques to measure and reduce the variability in the system of production. Relying on inspection to fight fires as they arise guarantees that you will continue to produce the same quality product you presently produce. SPC is a long-term philosophy, but it must begin with top-level management. Only when new questions are asked based upon an understanding of statistical thinking will all employees be free to work towards improving the quality of the products they produce.

If additional informations is reqested please feel free to contact us.

Fabricators and Manufacturers Association, Intl'
5411 E. State Street
Rockford IL   61108
815/399-8700

Presented at the SME Deburring and Surface Conditioning '87
Conference, February 1987

Quality Vibratory Finishing: Preparing for SPC

Samuel R. Thompson
Ultramatic Equipment Company

INTRODUCTION

Mechanical finishing is a profession, not simply a convenience
operation or necessary evil that a manufacturer must put up with. Yet,
we continue to find manufacturers who will up-date operations with CNC
equipment producing all parts alike, then deburr by hand, creating no
two parts alike. Too many manufacturers are operating without the
foggiest notion of deburring and surface conditioning theory; although
both are cornerstones of their profession.

While manufacturing becomes automated at a faster pace and cutting
tool manufacturers concentrate on metal removal, machine tool builders
on infinate precision, and production repeatability, the vibratory
finishing industry has not been sitting on their hands. Todays
vibratory systems are designed for processing precision parts and
quality production.

Where statistical process control (S.P.C.) methods are utilized
or being introduced they usually trickle down to the finishing area.
As with other machine tools; the vibratory process must be fully understood to make S.P.C. methods successful. S.P.C. demands documentation
of every aspect of the process enabling graphs or charts to be formulated plotting the objectives while overlaying the actual results.
Flucuations are immediately recognized allowing appropriate action or
decisions to be applied before excessive deviations from cost or
quality occur. Optomizing consistant quality at minimum cost.

This paper is written as an aid to understanding the mechanical flexibility of the equipment by application. The fundamentals of use, the selection of media and compound to obtain repeatable procedures compatable to S.P.C.

ADJUSTING THE VIBRATOR

All too frequently is the word amplitude misused in reference to vibratory equipment. Amplitude is the counter weight load on the vibrator shaft and has nothing to do with speed. The vibrator is the energy developing mechanism. It's rate of rotation is not affected by changing weights.

Tub type vibrators have only two adjustments - amplitude (weight) and vibrator RPM (if the machine is equipped with a variable speed device).

Most round bowl type vibrators have three adjustments. Amplitude, speed, and counterweight lead angle. (As in the tub equipment bowl machines must have a variable speed device to change the vibrator RPM.)

All bowl vibrators share a common factor: the vibrator is a verticle shaft arrangement located in the doughnut hole area of the bowl. Driven by a variety of methods, the shaft is weighted at the top and bottom. (Figure 1) The offset of these weights (flywheels) creates the eccentricity developing vibration. The relationship of these weights to one another (flywheel lead angle) (Figure 2) develops the mass motion or torridal path in the processing channel (similar to a coil spring incircled and connected at the ends). By changing the relationship of one flywheel to the other extends or closes the spiral of the mass path. (See Figure 2A) The energy development of the vibrator is governed by the amount of counterweight at the shaft ends. More weight equals heavy loads, less for light work.

VARIABLE SPEED

Speed (RPM) changes permit adjustments to the mass motion without stopping the machine to compliment the process ie/ part to media contact time, cycle timing, preventing part to part contact, controling separation time etc. Contrary to popular belief, changing or adjusting weights does not change anything other than amplitude and or lead angle. (See Adjusting The Vibrator) Usually, once set the fly weights are not changed unless the part or process changes. Not all machines have the ability to change speed, nor is it always required, yet it can be a valuable asset where a variety of parts are to be processed in the same machine.

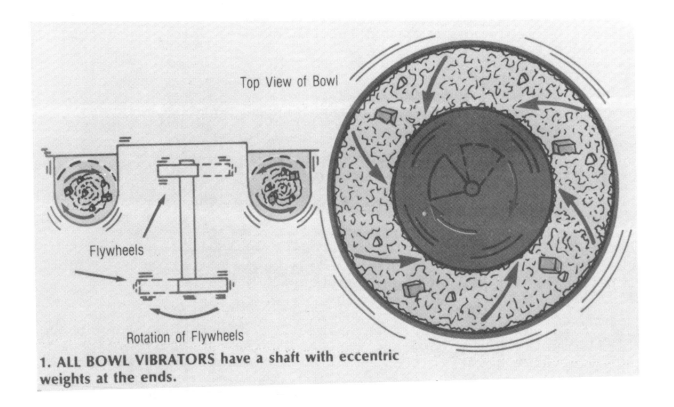

Top View of Bowl

Flywheels

Rotation of Flywheels

1. ALL BOWL VIBRATORS have a shaft with eccentric weights at the ends.

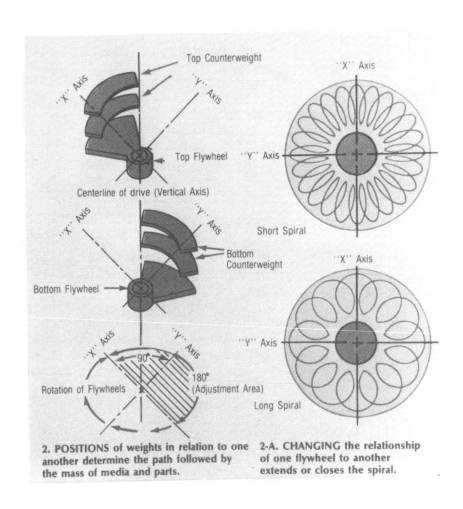

Top Counterweight

"X" Axis

"Y" Axis

Top Flywheel

"X" Axis

"Y" Axis

Centerline of drive (Vertical Axis)

Short Spiral

"X" Axis

"Y" Axis

Bottom Counterweight

Bottom Flywheel

"X" Axis

"Y" Axis

90°

180° (Adjustment Area)

Rotation of Flywheels

"X" Axis

"Y" Axis

Long Spiral

2. POSITIONS of weights in relation to one another determine the path followed by the mass of media and parts.

2-A. CHANGING the relationship of one flywheel to another extends or closes the spiral.

IN PREPARATION FOR ADJUSTMENTS

1. The processing channel must be full.
2. The compound must be flowing through properly, without excessive lubricity or foam.
3. The drains must be open and clear to accept the flow at a minimum rate of 1½ gph per cubic foot of processing capacity. Excessive liquid lessens the ability of the vibrator to move the mass properly. Too dry is just as bad. (Unless the process is intended to run as a dry process, using wood pegs, corn cob meal or similar materials.)

AMPLITUDE ADJUSTMENT (COUNTERWEIGHT)

Counterweight is the strength portion of the vibrator mechanism. More counterweight is required to process heavier loads, less for light work.

FLYWHEEL LEAD ANGLE ADJUSTMENT (BOWL MACHINES)

The flywheel relationship is changeable within a 180-degree adjustment. The maximum effective range in this area is approximately 150 degrees. By opening the flywheel relationship the spiral is opened, creating a long spiral extension. Close the relationship to shorten the spiral, creating more rotary motion. Lead-angle changes are made to compliment part size or shape as an aid to internal separation.

COMMON TUB AND CONTINUOUS VIBRATOR LOCATIONS

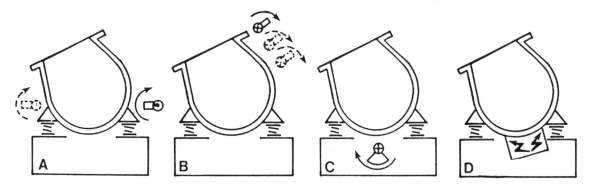

FIGURE 3

SIDE MOUNT DRIVE

ILLUSTRATION A

The side mount drive can either be single or double. The vibrator is located at the waistline of the tub between the upper and lower quadrant. When the double vibrator is employed, they are opposite one another front and back. They are synchronous and amplitude weight must be equal for each shaft.

"FIGURE 4   O.H.D. DOUBLE SHAFT VIBRATOR"

OVERHEAD DRIVE
ILLUSTRATION B

Introduced in 1974, the Overhead Drive (OHD) is the newest concept in vibrators. Since the vibrator is located above the springs and over the mass, there is no requirement to lift the tub to move the mass. The overhead design allows for multiple vibrators in one location. If more than one is required the shaft and flyweights will be synchronous, yet the amplitude weight load on each may be different to compliment the part/process or application.

UNDER TUB DRIVE
ILLUSTRATION C

The introduction of the tub type vibrators in the 1950's featured this method and it continues today. The vibrator is attached beneath the tub itself having pie-shaped or oblong flyweights fitted to the shaft. Vibration energy development is upward producing a rise and fall motion to the tub.

MAGNETIC DRIVE
ILLUSTRATION D

Machines equipped with magnetic vibrators are less than ten cubic feet in capacity. The vibration input is from the underside of the tub and operates at a constant 3,600 cycles per minute. Voltage is controlled via a rectifier console. Increasing the voltage strengthens the magnetic field resulting in a longer pull to the armiture, thus increasing stroke. The theory is that the variable stroke assimilates the variable speed and amplitude of a mechanical vibrator.

MEDIA

Often the question is asked "How long does the media last?". A reasonable response is "How long does a saw blade last?". For an S.P.C. operation an attrition calculation should be in place as part of the process. The calculation would be one of the several assembled on a spread sheet that documents all the elements of a process for a specific part. Included would be formula or detail of media to part ratio, compound flow rates, and amplitude settings, etc.

By accurately charting media attrition and make up intervals, the reasons of quality fluctuations can be quickly monitored. A reasonable method to accomplish this is to run a broken-in production load of media for a given length of time and compare the before and after weight. A control part can be processed and also compared. Then by comparing media loss with metal loss, a factor of efficiency can be established. By duplicating the procedure with other media shapes, sizes, bonds, or manufacture, direct comparisons can be calculated for cost and efficiency.

Media comparison analysis cannot be calculated with a new load of media. It first must be broken into a typical in-production variable size condition. Routinely, this will not begin before no less than one-third of the original new material has worn and has been replenished at appropriate intervals. In production uniform mass motion can only be preserved by maintaining a full processing channel. Quality assurance can only be maintained by replenishing worn media in small quantities at specific intervals predetermined by the S.P.C. data.

MEDIA MOBILITY

The selection of media is usually part dependant. However, there are certain other factors that bear consideration. Media mobility is critical to the vibratory process, since vibratory motion is developed from kinetic energy rather than positive displacement (tumbling).

The mobility factor is what keeps the parts and media in harmonious motion. Without the uniformity of motion the parts and media get out of step, increasing the potential for part to part contact. (See Table 1.)

Other factors of importance include matching media size and abrasiveness to the task. A small media size having a coarse abrasive grain will produce a finer finish than a larger one containing the same abrasive. The small media having less contact area, produces a close pattern, resulting in a finer finish. If a small media is in use performing a deburring operation, it is logical to employ a fast cutting abrasive blend improving the overall efficiency to the process time and production value.

The Black Art theory of ceramic media for ferrous parts and plastic media for non-ferrous metals is not quite correct, nor is any theory correct until you decide what is required of the media. Example: In the case of a titanium deburring operation, plastic media worked best because it was softer and lighter weight which allowed more part to media contact time than the ceramic, which appeared to bounce off.

**Table I. Vibratory Media Versatility
Processing Characteristics by Shape**

MEDIA SHAPE	LODGING POTENTIAL LO HI	MOBILITY	SURFACE CONTACT AREA	MEDIA SHAPE RETENTION	SURFACE REFINEMENT	PROCESS TIME 1 long 5 short	VERSATILITY
SPHERE		5 B	1	5	5	1	17
V-CUT CYLINDER		4	5	5	5	5	24
CYLINDER		4 C	2	4	5	3	18
CONE		5	2	5	5	3	20
PYRAMID		3	5	2 D	4	4	18
ELLIPTICAL		4	3	4	5	3	19
TRIANGLE		3 A	5	4	4	5 E	21
DIAMOND		3 A	4	4	4	4 E	19
TETRAHEDRONS		2	4	2	4	4	16
STAR SHAPES		2 A	3	3	4	4	16

Numerical Scoring: 1 = Low, 5 = High. Efficiency is not rated by a high accumulated total. The total score is an indication of the media's all around versatility.
A. The mobility factor is best in moderate sizes and when more evenly dimensioned (⅜" × 5/16"). Poor when thin and broad (⅛" × 1-¼")
B. Small sizes absorb energy, dampening mass activity. The lack of any flat surfaces further impedes the transmission of energy development, limiting the use of spherical media
C. Mobility is restricted when width to length ratio exceeds 2 to 3.5
D. Loses edges and becomes a tapered elliptical shape
E. Thin flat shapes finish faster than those more evenly dimensioned

THE COMPOUND SELECTION

The intended purpose of compound is to compliment the process, keeping the parts and media clean by continously flushing away debris as quickly as it is generated, preventing any build up of foam or other undesirable qualities that would impede the vibratory process.

Traditionally, compounds were blended for use with specific metals. The current trend is to "all metal" compounds having low or no foam qualities with manageable lubricity and excellent cleaning qualities. These new blends are employed for abrasive applications ranging from deburring to pre-plate finishing. By contrast "specific" compounds are very directed by application ie/ descaling, burnishing, or degreasing - they usually are in place as part of a system of continuing same requirement volume production.

Compound concentration is important in the area of an abrasive application, should the process be over compounded the excessive compound will form a barrier preventing the media from complete contact with the workpiece thereby decreasing it's efficiency. A simple test is to vigorously rub two fingers together under the

compound dispensing inlet until they become slippery. Then while
still under the flow, very lightly rub. If you can feel your finger-
print lines, the lubricity concentration is correct and not impeding
the media's intended abrading efficiency. Simply put, the entire
operation should be foam free, look good and clean, with a smooth
uniform mass motion.

COMPOUND MIX AND DISPENSING

Usually this is accomplished by an automatic system furnished by
the equipment manufacturer. Two systems are common to the industry,
each having specific benefits and intended use. Both dispense an
accurate amount of compound and are adjustable.

The preproportioning pump is commonly used one of two ways. Direct
injection into the water line or via a hose direct to the processing
channel. The pump functions via a piston which is variable in
function. Two controls govern it's activity - one for length of
stroke, the other for frequency of stroke. The graduated numbers on
the control are for reference relationship only. The percentage of
compound ratio is established by adjusting the desired flow of water,
then setting the pump by trial and error measuring, to dispense the
desired amount of compound for the process. Then marking the pump
settings for length of stroke and frequency. For example: the
measured consumption of compound was one gallon for fifty-five gallons
of water, approximating a two percent solution. Obviously, the system
requires documentation and locking in the settings prior to release
for production, however, the flexibility of the system is valuable
where a variety of media, compound, parts, and processes require
frequent change.

The venturi system is less complex to operate and is employed
where the equipment or process is more directed even though a variety
of parts may be processed or the media changed. The process (burnishing, deburring, pre-plate, etc.) would be the same.

The venturi system vacuums the compound through an orfice of a
diameter matching the compound viscosity via water pressure passing
by the venturi. A restrictor on the afterside of the venturi is incorporated in the drop tube and maintains a flow restriction that
develops the back pressure insuring a uniform mix. The uniqueness
of the mechanism is that it involves fewer moving parts and controls.
The mixed solution flows from the drop tube into a holding tank from
which it is pumped on demand to the machine. A float is connected to
the incoming water valve which automatically controls the make up
reserve of mixed solution. One feature of this system is once the
mix ratio orfice is installed the compound concentration mixed in
solution remains constant since there are no adjustments short of
changing orfice sizes. The only operator adjustment is the mixed
solution flow rate to the machine, the compound consumption is regulated by only that one variable. Other advantages are by simple
addition of pumps, other machines can be served by the same reservoir,
the reservoir can be remotely located convenient to the compound
storage area.

The importance of proper compounding cannot be overstated. Excessive lubricity (slipperyness) can play havoc with process time, excessive foam retains dirt and liquid which dampens the mass motion (properly adjusted machines do not need foam to cushion the amplitude). The machine drains should be clear, allowing an unrestricted flow. Flooding the mass dampens the activity. In contrast, insufficient flow causes a dirty condition, contaminating the parts and insulating the media from performing properly.

POWDER COMPOUND

Dry cleaning compounds are used effectively for tumbling processes and it is only natural that their use can carry over to the vibrator. However, each tool has it's use, and the application of dry compound to a vibratory installation transfers the scientific requirement of consistant repeatibility to the operators artistic ability to do the same thing every time all the time. Obviously, a procedure of this nature is not accurate and cannot be considered in a vibratory installation.

WASTE DISPOSAL AND RECIRCULATION

The immediate thought for recirculation of the liquid flow through cleaning solution in a vibrator is worthy of consideration, if the disposal of liquid waste on a continious basis will present a genuine problem. However, several points must be considered before proceeding.

Recirculating the liquid cleaning solution reintroduces dirty solution to the mass containing small particles which were removed from the parts and media during the previous cycle, quickly reducing them to contaminating liquified solids which redeposit upon the part. These solids being so fine will also deposit upon the media glazing it's surface reducing efficiency. The longer the liquid is recycled the more concentrated and contaminating it becomes. The obvious alternative is to capture the used liquid, renew it to it's original clear state then reuse it.

A vibratory process utilizing abrasive media generates solids on a continous basis no less than five percent of which are smaller than four microns, particles that small can be considered liquidified solids ie/ they remain suspended in the spent liquid for indefinate time (days). Plastic media being softer than ceramic media will generate a finer particle waste more quicly than ceramic. It's softness and lighter weight produces more media solids in suspension for a longer period of time. Filtering will not clean the liquid to a clear state which indicates solids remain along with some cleaning chemicals. The make up amount of chemical feed also becomes tricky since some chemical ingredients remain with the collected solids and others with the recirculated liquid.

Usually most smaller vibratory installations can only justify the cost of settling tanks - cyclonic friction cleaning - or simple

filteration. None of which clean the solution to it's original
clear state, yet may be adequate for certain product applications that
do not require precision repeatability or containment free parts.
Treatment systems are available that will adequately clean the spent
solution to clear neutral state for re-use however, such equipment
must be specified for each particular application to remain cost
effective while maintaining consistant quality standards.

Justification for any system obviously is only calculated against
alternate costs, so a ten cubic foot burnisher producing button hooks
four hours a day would never justify itself as fast as a twenty cubic
foot machine precision finishing bearing cages operating eight hours
a day.

The chemical cleaning compound used in the process can overcome any
system if it's compatibility is not matched to the method and maintained as designed. The slightest change potentially can clog filters
lubricate a cyclonic sytem sufficently to discharge solids rather
than collect them, create unmanageable excessive foam, cause oders,
and foul the system to a point that it becomes useless. In contrast,
any system will be user friendly as long as it's particular capabilities are fully understood and operated as such.

CONCLUSION

While there are a broad selection of documents available about
vibratory finishing, few document research statistics accurately
enough enabling an engineer to sort through finding the specific
variables necessary to establish S.P.C. to vibratory finishing. However, the noted references in this paper contain excellent documentation of methods comparison, a computer program and mechanical
explanation to initiate S.P.C. standards for most any situation.
Obviously, charts and overlays are self generated by standard S.P.C.
guidelines to suit your requirements.

REFERENCES

PART FACTORS	Size and Configuration	(9)
	Burr/Surface Condition	(2) (8)
	Material Finishability	(2)
	Quality Finish Requirement	(2) (4)
EQUIPMENT FACTORS	Mechanical Adjustments	(3) (7) (10)
	Machine Capabilities	(3) (4) (5) (8) (9) (10)
	Processing Channel Limitations	(3) (5) (6) (7) (10)
	Load Ratio	(2) (4) (5) (9)
MEDIA	Size and Shape	(5) (6)
	Type of Bond Abrasive Qualities	(1) (4) (6)
	Attrition Factor	(1) (2) (5) (8) (9)
PROCESSING	Cycle Time	(2) (6) (9)
	Compound Volume, Compound Lubricity	(4) (6) (9)

Material Handling		(3) (4) (10)
Method Examples		(2) (8)

COSTING   Application Examples   (2) (8) (9)

(1)   Jones, G.F. Testing and Selection of
        Vibratory Finishing Media SME Technical
        Paper M R 79-745

(2)   Gillespie, LaRoux K.
        Deburring Capabilities and Limitations
        Copywrite 1976 by The Society of Manufact-
        uring Engineers Dearborn, Michigan

(3)   Thompson, Samuel, R.
        Selection of Vibratory Finishing Equipment
        SME Technical Paper M R - 81 - 381

(4)   Kittredge, John B.
        Understanding Vibratory Finishing Parts
        1-4 Products Finishing, February - May 1981
        Single Pamphlet Reprint Available
        Kittredge Consultants, Kalamazoo, Michigan

(5)   Kittredge, John B.
        The Mathematics of Mass Finishing
        SME Technical Paper M R - 81- 399

(6)   Thompson, Samuel R.
        Shaping up Mass Finishing
        Metal Finishing Magazine April 1986
        Reprints Available
        Untramatic Equipment, Addison, Illinois

(7)   Thompson, Samuel R.
        Bowl-Type Vibrators
        Products Finishing Magazine September 1986
        Reprints Available
        Ultramatic Equipment, Addison, Illinois

(8)   Cost Guide For Automatic Finishing Processes
        Society of Manufacturing Engineers
        Dearborn, Michigan
        Edited by Lawrence S. Rhodes, Extrude Hone
        Corporation

(9)   M.F. Calc Computer Program
        by John B. Kittridge Consultants
        Kalamazoo, Michigan
        Order From SME or Direct From Author

(10)   Thompson, Samuel R.
            Comparison Of Retangular Tub and Inline
            Continuous Vibratory Equipment
            SME Technical Paper M R 83 - 679

                    SPECIAL THANKS TO

            METAL FINISHING MAGAZINE
                HACKENSACK, NJ

                        AND

            PRODUCTS FINISHING MAGAZINE
                CINCINNATI, OH

For the use of illustrations and information from
their publications.

Reprinted from *Modern Casting*, May and June 1983

Statistical Quality Control In the Foundry—Part 1

Part 1 of this two part feature on statistical quality control begins a discussion on the theory behind data sampling and the construction of standard deviation curves and X-bar-R charts.

Archibald Jamieson
Jamieson Management Services, Ltd
St. Catharines, Ontario, Canada

In the early part of this century, the fathers of scientific management were advocating radical changes in the ways in which work should be performed. They recommended the analysis of current procedures, the elimination of unnecessary movements, the rearranging of the remaining activities in their "best possible" sequence, the timing of the work and, finally, a payment of some share of the savings to the workers involved. This innovative approach brought about such remarkable increases in productivity in so many companies that the word soon got around and people began trying to copy the originators.

Unfortunately, there is a human tendency to latch onto things which can be seen faster than onto things as abstract as ideas or concepts. People had been seen watching workers before but the stopwatch was something new, so the time-study became the means of increasing productivity and the stopwatch became the tool whereby this was to be accomplished.

In all too many cases, however, stopwatches were given to people with no training, whatsoever, in their use, together with vague instructions such as, "Time how long it takes to make a mold on the squeezer machine." The end result of much of this was that piecework prices or bonus rates were established before the methods analysis and work simplification was done. The workers then proceeded to do their own methods analysis and work simplification which resulted in large increases in their earnings.

Panic followed when the earnings exceeded those of the foreman and time after time rates were cut to bring earnings back down again so that often the producer was working harder just to earn his standard wage. Is it any wonder that labor disputes followed and that timestudy became an unpopular practice with both management and unions in many plants? By adopting only part of a concept these companies had thrown the whole concept into disrepute.

As if this was not bad enough, companies then proceeded to make the same type of mistake when statistical quality control came along. Learned minds developed sampling theory and the ideas behind control charts for variables, but once again, these are concepts which are not seen or comprehended at a casual glance. What could be seen were the charts, so the charts became for statistical quality control what the stopwatch had become for work measurement.

Just as people were given stopwatches without training in either methods analysis or rating, people who had never heard of control limits or range started constructing charts. The author has vivid recollections of visiting the sand control laboratory of a fairly large international corporation, back in the late 1950s, and being shown "control charts" for sand properties. These charts had no mean (or aimed at values), no

control limits and, as individual values had been plotted, no range chart.

Such charts of individual values have such limited usefulness that it was only a matter of time before they, too, fell into disrepute and took with them the reputation of statistical controls. It had been done again. Just as the misapplication of

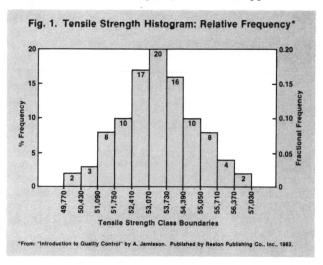

Fig. 1. Tensile Strength Histogram: Relative Frequency*

*From: "Introduction to Quality Control" by A. Jamieson. Published by Reston Publishing Co., Inc., 1982.

the stopwatch gave timestudy a bad name, the misapplication of charts gave statistical controls a bad name also. Statements such as, "You can't run a company with charts" gained their origin from such misapplication.

If unbiased consideration is to be given to the introduction of statistical controls, preconceived ideas which were probably developed through observation of improper or inadequate techniques must be put aside. What is more, there are certain issues which cannot be avoided. One must be prepared to take the time to understand some of the theory behind sampling if maximum benefit from the use of X-bar-R charts is to be obtained.

Variation in Numbers

If there is a desire to control the quality of an item or a process by measuring some variable then, as long as a sensitive enough method of measurement is used, there will be a variation in the results. This will result in one value which is the lowest of all, one which is the highest and many in between. The mean of these values can be found but it may not correspond to an actual measurement and may fall between two measured values.

Usually the frequency of occurrence in a range of values is not the same. There will not be as many occurrences at the low and high extremes as there are at or about the mean. Typical results will be distributed similar to those shown in Fig. 1. The ideal form of such a distribution of values is known as the normal distribution and is illustrated as a curve in Fig. 2.

An interesting characteristic of this ideal curve is that if the standard deviation is used as a measure of the amount of variation then some probability statements can be made about any particular range of values in a given system. (At this point, it is not absolutely essential to understand how standard deviation is calculated. It is sufficient to accept it as a useful measure of variation in a collection of numbers.)

From the mean to plus and minus one standard deviation, 68.27% of the values will exist; between plus and minus two standard deviations, 95.45%; and between plus and minus three standard deviations, 99.73%. This is useful information because plus and minus three standard deviations is selected as the range within which variation will be accepted as being "normal;" then there is only one chance in 370 of finding a value either above or below these limits. Variation within these limits will occur as a matter of chance and although the extremes will only be found occasionally, it should not be a surprise when they do occur. As can be seen from Fig. 2, three standard deviations is at the tailing-off of the curve and the probabilty of being even further beyond decreases at an extremely fast rate.

It is for this reason that plus and minus three standard deviations from the mean has been chosen as the range within which chance causes operate and chance variation occurs.

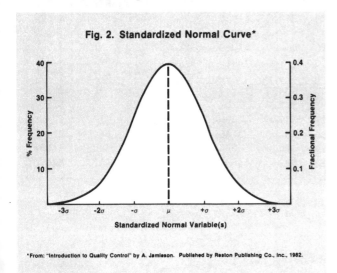

Fig. 2. Standardized Normal Curve*

*From: "Introduction to Quality Control" by A. Jamieson. Published by Reston Publishing Co., Inc., 1982.

The probability of measurements falling outside this range is so low that, if they do, it is certain that some particular event or condition brought this about and assignable causes can be looked for.

So, here is one of the first advantages in the use of statistical techniques: it permits the application of the *principles of management by exception* to the field of quality control. It gaurds against wasting time looking for the cause of a variation when simple chance was at work. Similarly, it imparts added confidence to continue searching when the probabilty of chance alone is so low as to make success in finding an assignable cause for a particular variation very likely.

Variation in Means

There is just one catch to all of what was said above and that is individual values often come close to normal but not close enough to be able to make probability statements with any reasonable degree of confidence. However, if individual values are grouped and the means (or arithmetic averages) found for each of these subgroups, then the mean values will come closer to being normally distributed than the individual values from which they were derived. Also, the larger the size of the subgroup the closer the means will come to normality. In fact, some distributions are twisted to one side (skewed) when looking at the individual values but become normal when when looked at as a distribution of the means of subgroups of these values.

When controlling a variable which is at, or close to, its maximum (or minimum) technical possibility, it is not uncommon to find a skewed distribution of values. There are two measures of central tendency which are of importance here: the mean and the mode. The mean is the arithmetic average; the mode is the most commonly occurring value; and, in a normal distribution, the two coincide.

However, if the removal of an impurity in a range, which is close to the minimum possible for the process in use, is to be controlled, it may be found that there are fewer values below the mode than above it. Conversely, if tensile strength of a metal in a range which is close to the maximum possible for the process in use is to be controlled, it may be found that there are fewer values above the mode than below it.

Table 1 contains 150 tensile strength results and illustrates the strength example. If the range of the highest to the lowest is divided into eight subdivisions or classes and the values in each class are counted, the histogram shown in Fig. 3 is obtained. The mean of these results is 53,020 psi but the mode occurs at 53,400 psi. Also, 46.7% of the values are below the mode and only 26.6% above it. This distribution is therefore

Table 1. Ultimate Tensile Strength of Class "50" Cast Iron

52080	52410	52080	52080	54060	53070	53730	52410	53730	51750
53400	52740	56040	53400	52740	52410	56040	49770	53070	51090
55050	53070	51750	53400	53730	55050	51750	52410	52080	51090
53070	52080	54720	52080	55710	53730	54390	51750	52410	52740
51420	55050	55710	52740	50100	52740	53070	55710	53730	50760
50760	55710	51420	55050	51750	53070	52410	53400	51090	54720
52080	53730	51090	53730	53070	55050	50100	53400	51090	52410
52410	53070	53400	50760	50430	55050	53730	51090	54720	54390
54060	52080	52740	52740	52410	54390	53400	52740	50100	52080
52410	51090	54720	51750	53730	51090	51090	52080	50760	52740
51090	55050	52410	50760	51750	54390	52080	53070	55050	54720
53730	53070	54720	51420	52740	51750	53730	51420	52080	52740
53400	52740	54720	53730	53070	53730	53400	53070	53070	54060
53070	54060	53730	52410	53730	52080	52740	53400	53070	54390
54060	54390	54390	54390	54060	51300	54390	54720	54060	54720

Fig. 3. Skewed Tensile Strength Distribution

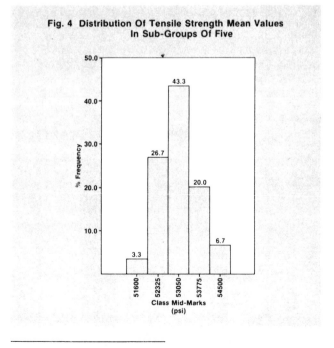

Class Mid-Marks (psi): 49440, 50430, 51420, 52410, 53400, 54390, 55380, 56370

% Frequency values: 0.7, 6.0, 14.7, 25.3, 26.7, 17.3, 8.0, 1.3

Table 2. Averages of Tensile Values in Sub-Groups of Five					
Sub-Group Number	Individual Values (x)				Sub-Group Mean (X̄)
1	52080	52410	52080	52080	54060	52500
2	53070	53730	52410	53730	51750	52900
3	53400	52740	56040	53400	52740	53700
4	52410	56040	49770	53070	51090	52500
5	55050	53070	51750	53400	53730	53400
6	55050	51750	52410	52080	51090	52500
7	53070	52080	54720	52080	55710	53500
8	53730	54390	51750	52410	52740	53000
9	51420	55050	55710	52740	50100	53000
10	52740	53070	55710	53730	50760	53200
11	50760	55710	51420	55050	51750	52900
12	53070	52410	53400	51090	54720	52900
13	52080	53730	51090	53730	53070	52700
14	55050	50100	53400	51090	52410	52400
15	52410	53070	53400	50760	50430	52000
16	55050	53730	51090	54720	54390	53800
17	54060	52080	52740	52740	52410	52800
18	54390	53400	52740	50100	52080	52500
19	52410	51090	54720	51750	53730	52700
20	51090	51090	52080	50760	52740	51600
21	51090	55050	52410	50760	51750	52200
22	54390	52080	53070	55050	54720	53900
23	53730	53070	54720	51420	52740	53100
24	51750	53730	51420	52080	52740	52300
25	53400	52740	54720	53730	53070	53500
26	53730	53400	53070	53070	54060	53500
27	53070	54060	53730	52410	53730	53400
28	52080	52740	53400	53070	54390	53100
29	54060	54390	54390	54390	54060	54300
30	54390	54390	54720	54060	54720	54500

skewed. Taking the values out of Table 1, and working from left to right along each row, they should be arranged in subgroups of five and the mean found for each subgroup. This arrangement is shown in Table 2. The lowest value is 51,600 psi. This compares with a low of 49,770, a high of 56,040 and a range of 6270 psi for the individual values.

Sampling always produces this effect; the range for the means is always less than the range for the individual values.

This is something that must be remembered later when interpreting control charts. Because there are only 30 means, compared to 150 individual values, these have been grouped into only five classes and are illustrated in Fig. 4. The mean and the mode coincide here but now there is only 30% of the values below the mode compared to 46.7% with the individual values. The number above the mode remains fairly constant at 26.7% and 26.6% respectively.

This trend toward normalizing is what takes place when the means are compared instead of individual values. A word of caution is necessary, however. The above example has been given for illustrative purposes only; it must not be thought of as a rigorous proof of sampling theory. For more empirical proof, larger collections of numbers would have to be used.

The measure of dispersion among the means is called the standard error of the means. It is given this name to distinguish it from the standard deviation but it is in fact the standard deviation of the means. One applies to individual values, the other to groups of values and there is a relationship between the two according to sampling theory. From observation of the two overall ranges, it might be expected that the standard deviation would be the larger of the two but it is unlikely that it would be expected that the difference would be the standard error of the means multiplied by the square root of the sample size.

$$\text{standard deviation} = \frac{}{(\text{standard error of the means}) \times \sqrt{\text{sample size}}}$$

So, if the standard error of the means is obtained from a large enough number of subgroups, an estimate of the standard deviation can be calculated. Notice that it is referred to as an estimate. This is because the standard deviation is supposed to describe a characteristic of the population of all values but in quality control work with variables, the entire population is rarely dealt with. Instead, more and more samples are accumulated.

If large enough subgroups are used and enough of them are collected, the cumulative mean of the samples will approach the mean of the population and the estimate of standard deviation will approach the standard deviation of the population as a whole. Estimate, then, is a word which must always be kept in mind when trying to go from facts to inferences.

Part 2 of this article, appearing in next month's MODERN CASTING, will continue the discussion on statistical quality control, taking a closer look at control charts.

Fig. 4 Distribution Of Tensile Strength Mean Values In Sub-Groups Of Five

Class Mid-Marks (psi): 51600, 52325, 53050, 53775, 54500

% Frequency values: 3.3, 26.7, 43.3, 20.0, 6.7

Archibald Jamieson is the author of **Introduction to Quality Control**, *published by Reston Publishing Co, Inc, Reston, VA 22090.*

Statistical Quality Control In the Foundry—Part 2

The conclusion of our two part article on statistical quality control takes the information obtained with standard deviation controls and X-bar-R charts, discussed in Part 1, and explains how to construct control charts for monitoring a variety of systems and functions. Also given is a range of guidelines which are necessary when initiating statistical controls.

Archibald Jamieson
Jamieson Management Services, Ltd
St. Catharines, Ontario, Canada

As discussed in Part 1 of this article, by collecting test results or measurements and putting them into subgroups we are able to make a probability statement about the product or process and determine whether or not the variation which is taking place is due to chance or to some more identifiable cause. However, with the use of charts we can get an instant picture of results over a fairly long period of time and are able to see if there is any trend toward a change in the mean, any cyclical characteristic to the measurements and whether or not we have experienced any out-of-control conditions.

Control Charts

If one more step is taken with the collection of tensile strength results, a table can be constructed which contains everything that is needed to set up an X-bar-R chart. At each subgroup the lowest value is subtracted from the highest to obtain the subgroup range. This is illustrated in Table 3. Range is related to standard error of the means by a constant factor dependent upon the subgroup size and these factors are listed under A_2 in Table 4. By means of this table the control limits for the mean and range charts can be quickly found.

Just as 99.73% of the time individual values will fall within the mean $(\overline{X})$ plus and minus three standard deviations in a normal distribution, so also 99.73% of the time the mean values will fall within the mean of the means $(\overline{\overline{X}})$ plus and minus three standard errors of the mean. The A_2 factor when multiplied by the mean of the ranges $(\overline{R})$ results in three standard errors of the mean and this is where control limits are set on the $\overline{X}$-chart. In a similar way D_3 and D_4 give the control limits for range but this is a skewed distribution so in this case the limits are not equally spaced from the mean.

In the example, the sample (or subgroup) size (n) is five, so from the table we find that the factors to use are:

$$A_2 = 0.577, \ D_3 = 0, \ D_4 = 2.114$$

The calculations then become:

$$\text{Sum of means} = 1,590,300$$
$$\text{Mean of means } (\overline{\overline{X}}) = 1,590,300 \div 30 = 53,010 \text{ psi}$$
$$\text{Sum of ranges} = 93,390$$
$$\text{Mean of ranges} = 93,390 \div 30 = 3110 \text{ psi}$$
$$\text{LCL}_{\overline{X}} = 53,010 - (0.577 \times 3110) = 53,010 - 1790 = 51,220 \text{ psi}$$
$$\text{UCL}_{\overline{X}} = 53,010 + 1790 = 54,800 \text{ psi}$$
$$\text{LCL}_R = 0 \times 3110 = 0$$
$$\text{UCL}_R = 2.114 \times 3110 = 6575 \text{ psi}$$

These results are then used to construct the combined control chart shown in Fig. 5. None of the values fall outside the control limits so it can be said that the results are within

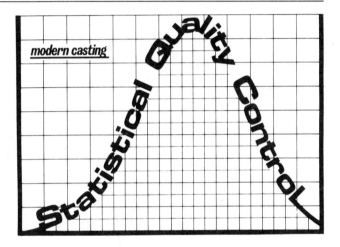

statistical control but care should be taken to ensure that a distinction is made between statistical control and technical control. A mean or range result is statistically out of control if it falls outside control limits but an individual result is technically out of control if it falls below a minimum specification or above a maximum specification. This is not apparent directly from the X-bar-R chart but can be deduced from the chart information.

The mean limits are set at plus and minus three standard errors of the mean so $A_2\overline{R}$ is equal to three standard errors of the mean. Sampling theory gave us the relationship between standard deviation and standard error of the mean so that three standard deviations $= A_2\overline{R}\sqrt{N}$. This gives an estimate about the population as a whole which can be called the process capability. Process capability $= \overline{\overline{X}} \pm$ three standard deviations and as has been stated previously, in a normal distribution this will contain 99.73% of all values. For the tensile strength example the calculations become:

$$\text{Process capability} = 53,010 \pm (1790\sqrt{5})$$
$$= 53,010 \pm 4003$$
$$= 49,010 \text{ psi to } 57,013 \text{ psi}$$

This distribution is negatively skewed, so the top end of the process capability may tend to be higher than that found in practice. At the low end of this distribution it is a different matter and attention must be paid to what it depicts. The low end depicts that, by chance alone, some of the time the

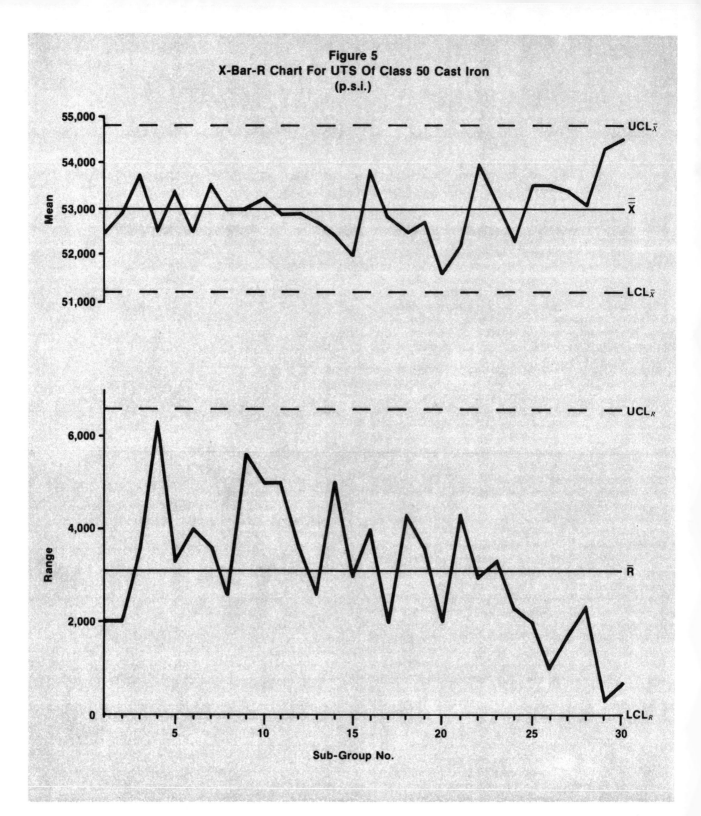

Figure 5
X-Bar-R Chart For UTS Of Class 50 Cast Iron
(p.s.i.)

strength will fall below the minimum specification of 50,000 psi. This is important information, because even if it had been noticed that the third value in the fourth subgroup was below 50,000 psi, it could be fairly certain that, in the long run, it was likely to happen.

Even if the 49,770 value had not occurred the results would have come out the same and it would have been possible to make the statement about the likelihood of being off-specification. Under these circumstances it would be necessary to make process changes so as to either increase the mean value (which may be difficult), or to tighten up all phases of the manufacturing process in an attempt to reduce the amount of variation.

Summary

By collecting test results or measurements in subgroups it is possible to make probability statements about the product or process and determine whether or not the variation which is taking place is due to chance or to some more identifiable cause. With the use of charts it is possible to get an instant picture of results over a fairly long period of time and be able to see if there is any trend toward a change in the mean, any cyclical characteristic to the measurements and whether or not any out-of-control conditions have been experienced. Only a chart will give this information at a glance.

One of the concerns that is sometimes heard with regards to charts is that they take too much time to keep up to date. This

Table 3							
X̄-R Record Sheet							
Sub-Group Number	U.T.S. Results (psi)					Mean (X̄)	Range (R)
1	52080	52410	52080	52080	54060	52500	1980
2	53070	53730	52410	53730	51750	52900	1980
3	53400	52740	56040	53400	52740	53700	3300
4	52410	56040	49770	53070	51090	53500	6270
5	55050	53070	51750	53400	53730	53400	3300
6	55050	51750	52410	52080	51090	52500	3960
7	53070	52080	54720	52080	55710	53500	3630
8	53730	54390	51750	52410	52740	53000	2640
9	51420	55050	55710	52740	50100	53000	5610
10	52740	53070	55710	53730	50760	53200	4950
11	50760	55710	51420	55050	51750	52900	4950
12	53070	52410	53400	51090	54720	52900	3630
13	52080	53730	51090	53730	53730	52700	2640
14	55050	50100	53400	51090	53070	52700	4950
15	52410	53070	53400	50760	50430	52000	2970
16	55050	53730	51090	54720	54390	53800	3960
17	54060	52080	52740	52740	52410	52800	1980
18	54390	53400	52740	50100	52080	52500	4290
19	52410	51090	54720	51750	53730	52700	3630
20	51090	51090	52080	50760	52740	51600	1980
21	51090	55050	52410	50760	51750	52200	4290
22	54390	52080	53070	55050	54720	53900	2970
23	53730	53070	54720	51420	52740	53100	3300
24	51750	53730	51420	52080	52740	52300	2310
25	53400	52740	54720	53730	53070	53500	1980
26	53730	53400	53070	53070	54060	53500	990
27	53070	54060	53730	52410	53730	53400	1650
28	52080	52740	53400	53070	54390	53100	2310
29	54060	54390	54390	54390	54060	54300	330
30	54390	54390	54720	54060	54720	54500	660

TABLE 4
FACTORS FOR CONTROL CHARTS

Sample Size n	Factor for:		
	Average	Range	
	A_2	D_3	d_4
2	1.880	0.0	3.268
3	1.023	0.0	2.574
4	0.729	0.0	2.282
5	0.577	0.0	2.114
6	0.483	0.0	2.004
7	0.419	0.076	1.924
8	0.373	0.136	1.864
9	0.337	0.184	1.816
10	0.308	0.223	1.777
11	0.285	0.256	1.744
12	0.266	0.284	1.717
13	0.249	0.308	1.692
14	0.235	0.329	1.671
15	0.223	0.348	1.652

Formulas for computing control limits:

$$UCL_X = X + A_2R$$
$$LCL_X = X - A_2R$$
$$UCL_R = D_4R$$
$$LCL_R = D_3R$$

is often nonsense and sometimes reveals a reluctance on the part of someone to publicize their own errors. Sometimes it reveals the fact that there are just plainly and simply too many charts. What does take up a lot of time is the initial processing of historical data in order to start up the first set of control charts but this is now being made easier by the use of microcomputers.

Because of the concerns, some guidelines are necessary when starting up statistical controls.

● Do not start with charts on everything and then cut back when they become a burden. Start with those products or materials which are produced to a specification. They are a must, but plot only the property or element of composition which has to meet the specification. When the advantages become apparent and experience is gained, then go to those compositional factors which critically affect the ability to meet the specification. Examples here might be the magnesium content of ductile iron or the total carbon content of a high-strength cast iron.

● Molding and core sand properties also play a critical role in the production of the final item and should be included in a program to introduce statistical controls. A similar approach to the one explained above should be taken. Experience has usually revealed that certain properties are either more difficult to control than others or that they are critical to the quality of the final casting. These are the ones which should be charted.

● Have a technical person make the decision about subgroup size and do all the periodic checks of control limits and process capability. Give the daily chart maintenance to a nontechnical person who would be able to drop everything they were doing if it became necessary to notify someone that an out-of-control condition had occurred. Speed is of the utmost importance when we have to look for an assignable cause of variation.

● Do not attempt to draw a chart with the precision of a draftsperson. They are neither engineering drawings nor works of art. Make them attractive to the eye however, so that they catch one's attention. Be sure to circle any out-of-control points and mark them as such. This is a form of management by exception and we must never run the risk of being accused of trying to cover up the exceptions.

The amount of variation in the range chart is directly related to the subgroup size so the latter should be clearly stated in the chart heading. Also, we must never cover up large variations by the use of a smaller scale. Both the X̄ and the R chart should be drawn to the same scale.

● Display the charts in a place in which anyone who is remotely interested can see them without having to ask for permission. Many people want to know the outcome of their work and charts can sometimes put meaning into an otherwise monotonous, routine job. Just make sure that someone takes the time to explain them and, particularly, that they do not represent individual results.

Finally, when we have enough information to feel confident that the mean of the means has stabilized, we can estimate the process capability for each of our quality characteristics and if a specification exists, check on the probability of meeting that specification. An example of a skewed distribution was used in this article but not all variables behave like this. In fact, most variables are normally distributed so we can have even greater confidence in the inferences which we make from our results.

CHAPTER 4

QUALITY CONTROL
IN THE AUTOMATED FACTORY

Presented at the SME Automation of Paint Lines Conference, May 1988

Statistical Process Control in an Automated Paint Facility

Jay L. Butler

Navistar International Corporation

There is increasing awareness on the part of the American manufacturer to recognize the need to incorporate Statistical Process Control (SPC) in manufacturing facilities. As automation increases in our plants, appropriate methods of statistical monitoring are sought. This paper offers insight into the strategies and potential but avoidable pitfalls in designing a system for SPC in an automated paint facility. The by focuses on insight gained through Navistar's development and implementation of a strategy to provide SPC and statistical product quality monitoring in its new, highly automated paint facility.

THE NAVISTAR AUTOMATED PAINT FACILITY

Navistar is in the process of opening a highly automated paint facility in Springfield, Ohio. Incorporated in the design of this facility is the ability to statistically monitor both the elements of the manufacturing process and the quality of the product being produced.

The new Navistar Paint Facility is a 210,000 square foot plant that is now in a start up mode. When the plant is fully operational, it will be capable of producing 1068 skids of material per day on a two shift basis. The substrata material of the various products painted are combinations of fiberglass, aluminum, and galvanized metals. The basic elements of the production system are two parallel pretreat lines, followed by immersion primer coating, body sealing, and finally robotically applied top coat. Each on-floor system module is controlled by Allen Bradely PLC´s communicating with a Vax host computer that monitors the entire operation. Interspersed in the production process are state-of-the-art ovens, manually applied interior coat, and twenty four robots in six separate booths for top coat paint application. Navistar is applying both thermal setting acrylic (TSA) and Base Coat/Clear Coat (BC/CC) in the facility.

DESIGNING THE STATISTICAL SYSTEM

At the onset, as you begin to plan your statistical system, it is important that you carefully determine what it is that you wish to accomplish. An automated facility creates an opportunity to gather an immense amount of process information in quantities heretofore unimaginable to American manufacturing. This virtual plethora of data will sorely tempt the statistically inclined to "WANT IT ALL". This temptation is particularly strong in companies just implementing statistical process control. (See figure #1)

In Navistar's case, there are thousands of sensors collecting data in micro second intervals. Each sensor is delivering information to on-floor programmable logic computers (PLC's), who in turn, report or communicate to a host computer that is monitoring the overall manufacturing process. (See figure #2)

One soon learns that while the software may be capable of handling this immense abundance of data, the hardware requirements required to archive and retrieve this data will seem to increase in exponential proportion to the data gathered. Therefore, close examination of the data points that the current or planned facility will be capturing is the first order of business.

Today's managers are interested in monitoring two types of data: Process and Product (See figure 3). Process data may be defined as data controlling the production process and environment in which the product is produced. Conversely, Product data provides information at the component serial number level about the specific product being produced.

Both are essential; one for information on the Process that is producing the product and the other about the individual product or component that is being produced. A Statistical "Process" related question might be: "WHAT HAS BEEN THE AVERAGE AND RANGE OF TEMPERATURE FOR ZONE 3 OF OVEN NUMBER TWO OVER THE PAST TWENTY DAYS?".

A typical "Product" related statistical question might be: "WHAT IS THE AVERAGE REJECTS PER HOUR ON CABS PRODUCED IN THE PAST TWENTY WORKING DAYS?".

SPECIFICATIONS FOR STATISTICAL SOFTWARE INTEGRATION

The importance of a clear, precise definition of system performance specifications cannot be overstated! Most problems can be traced to a period when the contractor and the customer each "Thought" they understood what the other was saying! Extra time spent in the planning phase of the program will pay off tenfold as you get further into system implementation.

DATA POINT SELECTION

Statisticians speak of variable and attribute data. To the statistician, attribute data is data that does not require actual measurement. Variable data, on the other hand, has a numeric value and can be measured. Typical variable data in an automated paint facility might be temperature, water ph, etc.

Software systems people, on the other hand, will speak of types of available information as either digital or analog data. In reality, they are speaking of the same thing as the statistician. (See figure #4) Sensor points in an automated facility that capture data such as "SWITCH OFF/ON" are known as "DIGITAL" sensors. These, the statistician would call "ATTRIBUTE" type sensors. A "DIGITAL" (Attribute) sensor may monitor whether a booth damper is open or closed. This particular damper must be open if the system is to operate correctly. While this information is immensely important to the successful operation of the facility, there is little need to statistically monitor how often the switch is in the "OPEN" position.

At the same time, another sensor in the same booth may be capturing the booth ambient temperature. This will be an "ANALOG" sensor and will be providing numeric data that is of statistical value to monitoring the performance of the facility. For example, it is important to know if the average temperature (x-bar) of an oven has been creeping upward over a period of time, even though it is still within the system alarm points. (See figure #5)

If the operating system is still on the drawing board, these necessary analog sensor points can be designed in. Careful review is necessary at this point because sensors can be added at minimal cost, if added as the system is being built. Adding these sensors later can be accomplished, but added software modifications are often needed in addition to the associated hardware costs. As you review the system, continue to ask yourself, "WHAT INFORMATION MUST I BE AWARE OF TO ASSURE MEETING BOTH OUR PRODUCT SPECIFICATIONS AND OUR OVERALL SYSTEM EXPECTATIONS?" It is advisable to speak to the current paint production personnel to be sure that one understands what parameters are important.

Do not assume that the contractors are statistically literate After all, they are going through the same statistical reawakening as are the OEM truck and automotive manufacturers. I have mentioned the relationship of attribute/variable and analog/digital information. It is very important that you review all system data points item by item with your contractors. Your statistical expectations and the necessary data points should be clearly defined in your contract specifications. Contract language will be discussed in greater detail further in this paper.

SELECTING STATISTICAL REPORT FORMATS

An important step that should not be overlooked is the selection of desired statistical report formats. The more advanced statistician will be familiar with many statistical charting methods, tables, and associated formulas. While these are excellent tools for the experienced, it is not what we are looking for on the shop floor. What is needed is a statistical reporting format that provides the necessary information to run the facility and does not require either advanced statistical or computer training to use.

In other words, all that is probably needed is software that will plot x-bar and range charts, and perhaps histograms. (See Figure #6) For the slightly more advanced, CP and CPK INDEX calculations should also be included. The Golden Rule is: "PROVIDE OUR PEOPLE WITH THE TOOLS NEEDED TO RUN THEIR BUSINESS." A sure way to make sure that they never use the statistical tools that we have put in their tool box is to make them so complex that they have to stop and take a refresher course just to understand what the chart is saying!

SELECTION OF SOFTWARE

Ideally, the software should be cursor driven and require only minimal familiarity with the access key board. This reduces any element of fear and uncertainty on the part of the casual user. A suggested cursor driven menu might look like this:

WHICH SYSTEM?	WHAT TIME PERIOD	CHART
PRE-TREAT LINE E-COAT LINE E-COAT OVEN INTERIOR OVEN TOP COAT BOOTH TOP COAT OVEN	THIS SHIFT MOST RECENT 8 HR LAST 5 DAYS LAST 20 DAYS	GRAPH ONLY? DATA ONLY? BOTH?
	WHICH ELEMENT?	WHICH SPC GRAPH?
WHICH LINE?	HUMIDITY TEMPERATURE	X-BAR & R HISTOGRAM
LINE ONE LINE TWO		
WHICH PROCESS?		
ZONE #1 ZONE #2 ZONE #3		

As you begin to work with your contractor, there are

several approaches that might be considered in selecting the appropriate statistical software. A basic requirement should be "EASE OF USE". Ideally, it should not require computer literacy in order to get information from the system. This can be achieved by a cursor/menu driven system that walks the user thru each of the necessary steps.

An approach might be to give a precise definition of your expectations to the software/hardware contractor and then allow him to develop the statistical software from scratch. This is the surest way to get exactly what you want. It is also often the most expensive and will generally result in low-resolution graphics that are functional, but not presentation perfect.

A second option is to accept shelf software provided by the prime contractor. Again, this will work, but will often require additional "COMPUTER LITERACY" on the part of the shop floor personnel. You are also restricted to the operating parameters of the shelf software selected.

A third, and recommended option, is modified shelf software. Many software companies are now offering statistical packages that allow insertion of "MACRO" programs at the beginning of their programs to assist in downloading. This option allows the contractor to write software interfaces that bridge his software to the selected statistical software. With this approach, you gain the benefit of some custom tailoring as well as high resolution graphics.

Regardless of the approach ultimately taken, care should be taken to assure that it is the prime contractors responsibility to assure that the system works and performs as specified.

DATA CAPTURE FREQUENCY

As mentioned earlier, the on-floor sensors are monitoring and reporting data continuously. Typically, the on-floor PLC's are collecting this information and forwarding it to the host computer at assigned time intervals. It is important that you assure that you can preset a number of individual sensor reports into a single chart plot point selected at the option of facility operating personnel. For example, a particular sensor might be forwarding information every 60 seconds. If these cannot be grouped and averaged, a graph for a single 8 hour production period would have 480 data points. (See figure #7) Over a 20 day operating month, the number of data points would climb to 9600 points on the chart.

DATA BASE HISTORICAL UPDATING

Care should be taken to specifically define the time periods for which you would like to download and plot information. Typical options desired might be: REAL TIME, NEAR-REAL-TIME, AND BATCH LOADING. Real Time is the most advantageous, but causes additional software work and considerably more storage capacity. As a result, this can be considerably more expensive.

NEAR-REAL-TIME provides information as recent as perhaps the last fifteen minutes. For day to day operations, this is as "REAL TIME" as is necessary and can be achieved for less expense.

The third option, and least acceptable is BATCH UPDATING, which usually occurs during off shift hours. In the author's opinion, this is unacceptable, as it does not allow the production personnel to statistically observe their performance during the production shift.

HISTORICAL RECORD RETENTION

In order for statistical process analysis to achieve its full potential, it is necessary that you look at data from a historical as well as a near-real-time vantage point. Your specifications should require that the on-floor access of data should be immediately available for the past 20 working days without the computer room downloading from a archived data spool.

Typical time periods that might be available on a good access menu might be:

THIS SHIFT
MOST RECENT 8 HOURS
PAST FIVE DAYS
PAST 20 DAYS

THE BID SPECIFICATIONS

The contract bid specification should include the following requirements:

"THE SOFTWARE SHALL PROVIDE A STATISTICAL ANALYSIS AND TRENDING PACKAGE FOR USE BY PLANT PERSONNEL. THE PACKAGE SHALL BE CAPABLE OF ACCESSING ALL SPECIFIED DATA ON THE SYSTEM AS WELL AS RECALLING ARCHIVED DATA. THE SYSTEM

REQUIREMENTS ARE CATEGORIZED IN THE FOLLOWING
SECTIONS":

(DETAIL YOUR REQUIREMENTS BASED ON ABOVE DATA)

TRAINING REQUIREMENTS

When writing your specifications, you should consider
what might be needed in order to use the statistical package
as an internal part of the plant operating software.
Training in statistical concepts and understanding of x-bar
charts should be considered a separate "NEED" and handled
independently of the contract with the software/hardware
contractor.

SUMMARY

With careful planning, it is possible to design a system
that will provide statistical process and product control for
the automated paint application plant. Designed correctly,
it will be a tool used daily by all plant personnel and not
just a device for use by the "QUALITY" group.

The ideal time to design the system is with initial
construction. This should not, however, prevent the addition
of a "STATISTICAL" package as an add-on to existing
software/hardware in an established facility.

It is the hope of the author that some of the lessons
learned by Navistar may be utilized by other OEM manufactures
as American industry moves toward increased statistical
literacy and expanded use of statistics in the automated
facilities now being built in the United States.

BIBLIOGRAPHY

Huntsberger, David V., Billingsley, Patrick ELEMENTS OF
STATISTICAL INFERENCE Allyn and Bacon 1979

Grant, Eugene L., Leavenworth, Richard S. STATISTICAL QUALITY
CONTROL Mc Graw-Hill 1974

Scherkenbach, William W. THE DEMING ROUTE TO QUALITY AND
PRODUCTIVITY Creep Press 1986

Mann, Nancy R. PHD., THE KEYS TO EXCELLENCE Prestwick Books
1985

AUTOMATION OF PAINT LINES
SPC IN AUTOMATED PAINT FACILITY

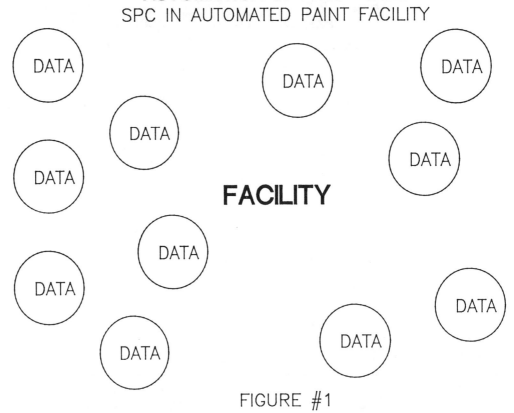

FIGURE #1

AUTOMATION OF PAINT LINES
SPC IN AUTOMATED PAINT FACILITY

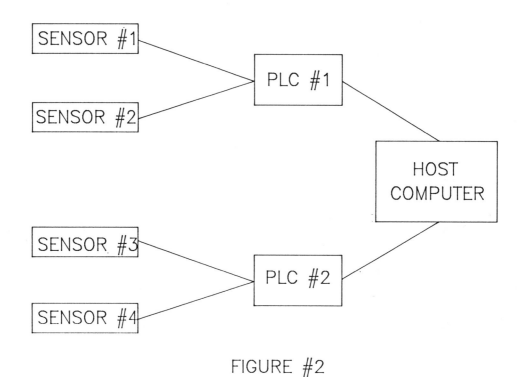

FIGURE #2

AUTOMATION OF PAINT LINES
SPC IN AUTOMATED PAINT FACILITY

TODAY'S MANAGER

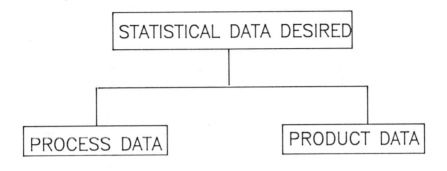

FIGURE #3

AUTOMATION OF PAINT LINES
SPC IN AUTOMATED PAINT FACILITY

| VARIABLE | = | ANALOG | = | OVEN TEMPERATURE |

| ATTRIBUTE | = | DIGITAL | = | OFF/ON |

FIGURE #4

AUTOMATION OF PAINT LINES
SPC IN AUTOMATED PAINT FACILITY

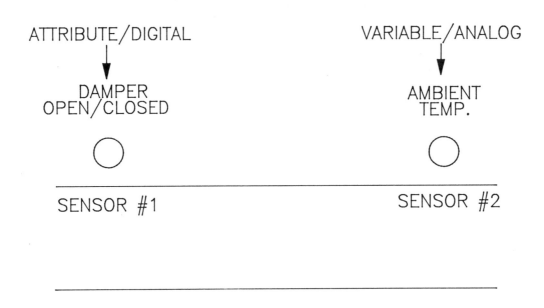

ATTRIBUTE/DIGITAL

DAMPER
OPEN/CLOSED

○

SENSOR #1

VARIABLE/ANALOG

AMBIENT
TEMP.

○

SENSOR #2

FIGURE #5

AUTOMATION OF PAINT LINES
SPC IN AUTOMATED PAINT FACILITY

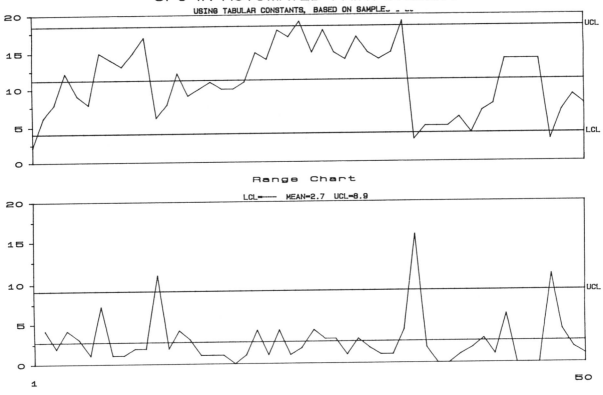

USING TABULAR CONSTANTS, BASED ON SAMPLE. . ..

Range Chart

LCL▬▬▬ MEAN=2.7 UCL=8.9

FIGURE #6

AUTOMATION OF PAINT LINES
SPC IN AUTOMATED PAINT FACILITY

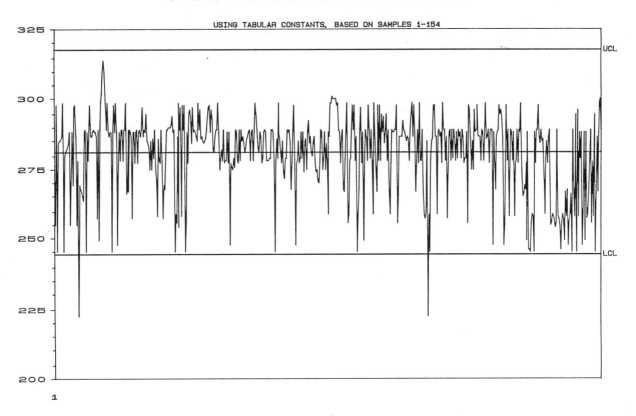

USING TABULAR CONSTANTS, BASED ON SAMPLES 1-154

FIGURE #7

Presented at the SME ROBOTS 11/17th ISIR Conference, April 1987

On-Line Statistical Process Control

Michael C. Barrett

IBM Corporation
Austin, Texas

A compiled BASIC program plots process control charts for six key robot and tool performance parameters on a graphics monitor tied to an IBM PC/XT station controller. Performance data are derived from status and error messages from the GMF Model MIA robot and a relay logic control program running concurrently in the PC. The process average is set by the user, and control limits are calculated from the data. The program plots the control charts while it is accepting messages and monitoring the cell status. The graph is current through the last machine completed.

INTRODUCTION

The robot station of interest is part of an automated assembly line. The station consists of a GMF (1) Model M1A robot, a horizontal axis dual screwdriver, a bushing inserter, a vertical axis screwdriver (stud installer), a conveyor system, and the computers that comprise the control system.

The control system includes the GMF Model C robot controller and an IBM PC/XT (2) station controller. A Texas · Instruments Model TI128 Programmable Logic Controller (PLC) is included in the dual screwdriver, but its only function is to control the operation of the tool.

The performance of the station is monitored by a portion of the station controller program that decodes status and event messages sent to it by the internal PLC and the robot. The messages allow the program to collect performance data related to the assembly operation.

(1) GMF is a trademark of GMF Robotics Corporation.
(2) IBM PC/XT is a trademark of International Business Machines Corporation.

CONTROL SYSTEM

Figure 1 shows the interconnections of the control components. The PC/XT includes a PLC system that interrupts the PC at 110 ms intervals. Operation of the PLC is transparent to the program that is running on the PC at the time. The ASYNC operation is also transparent since incoming data is stored in a buffer until it is retrieved by the operating program.

A BASIC program operates in the foreground as the robot is running. All communications with the outside world are handled by this program. The program also receives and processes information from the robot and the tools.

The incoming messages may generate outputs, increment or decrement counters, or start or stop timers. Information generated by the counters and timers is saved in files on the hard disk. The information is stored in such a way that it may be retrieved and graphed on the PC monitor.

Figure 2 shows portions of the message and data files that are maintained for the system being described. (NOTE: While the general tendencies of the data are correct, all the data presented in this report are simulated.)

BASIC CONTROL PROGRAM

Operator Interface

The operator receives information from the graphics monitor and gives information through the keyboard. The operator calls up various sub menus from the main menu. The main menu includes an option to display statistical process control charts. When this option is selected another menu appears that allows the operator to select which graphs he wishes to see.

PLC Interface

The PLC interfaces with the BASIC program through a single dimension integer array with 140 elements. Any of the 140 variables having a non-zero value represents a message or signal being sent from the PLC to the BASIC program. In some cases the value of the variable is reset by the BASIC program, and in other cases it is reset by the PLC.

The PLC interfaces with the robot and tools through conventional 24 volt, solid state relays.

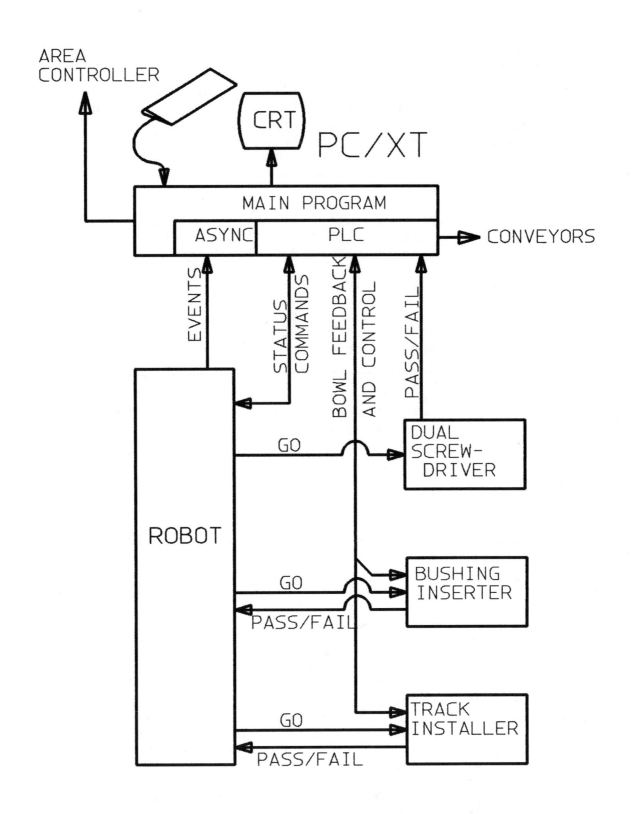

FIGURE 1 - ROBOT STATION CONTROL STRUCTURE.

```
04-01-1980  12:35:13      GM211  RSR-2 EXECUTED - ASM STARTED
04-01-1980  12:35:30      GM951  ASM'D REAR CAGE TAKEN FROM RACK-POS 71
04-01-1980  12:36:02      GM247  REAR CAGE PLACED IN PRODUCT
04-01-1980  12:36:14      GM910  FRONT CAGE TAKEN FROM BOX-POS 4
04-01-1980  12:37:09      GM255  TRACKS INSTALLED IN FRONT CAGE
04-01-1980  12:37:47      GM249  FRONT CAGE PLACED IN PRODUCT
04-01-1980  12:37:56      GM110  NEW BOX OF FEET REQ'D
04-01-1980  12:38:05      GM256  FOOT PLACED ON PRODUCT
04-01-1980  12:38:22      LD102  FOOT SCREWS DRIVEN (GOOD PART)
```

A.   Message file.   (Sequential file.)

```
DATE          TIME        OPERATOR INTERVENTIONS
 :             :          :DSD FAILURES
 :             :          ::DSD RETRIES
 :             :          :::FAR SIDE DSD FAILURES
 :             :          ::::SLOTTED HOLE DSD FAILURES
 :             :          :::::  BUILD TIME
 :             :          :::::  :     OPERATOR INTERVENTION TIME
 :             :          :::::  :     : LINE END (CR,LF,":")
 :             :          :::::  :     : :
 v             v          vvvvv  v     v v
04-01-198011:33:0000000   22          :
04-01-198011:34:5400000   114         :
04-01-198011:35:1410000   19    95:
```

B.   Data file GRSBLDTM.RAM, containing build time, operator
     intervention, and dual screwdriver failure data by product.
     (Random access file. Note the carriage return,line feed,
     and ":" to allow file editing.)

```
DATE     TIME   TOTAL BUSHING INSERTION ATTEMPTS
 :        :     : TOTAL BUSHING INSERTION FAILURES
 :        :     : : REAR CAGE ATTEMPTS
 :        :     : : : REAR CAGE FAILURES
 :        :     : : : : FRONT CAGE ATTEMPTS
 :        :     : : : : : FRONT CAGE FAILURES
 :        :     : : : : : :              :FRONT CAGE :            :FRONT CAGE :
 :        :     : : : : : :    REAR CAGE :FAILURES :   REAR CAGE  : ATTEMPTS :
 :        :     : : : : : :   FAILURES BY HOLE : BY HOLE :  ATTEMPTS BY HOLE : BY HOLE :
 :        :     : : : : : : 1 2 3 4 5 6 7 8 1 2 3 4 1 2 3 4 5 6 7 8 1 2 3 4
 :        :     : : : : : : : : : : : : : : : : : : : : : : : : : : : : : : :
04-01-198018:20 102 2 60 1 42 1 0 0 0 1 0 0 0 0 0 1 0 0 8 8 8 8 7 7 7 7 11 11 10 10:
04-01-198018:26 100 2 88 0 20 2 0 0 0 0 0 0 0 0 1 0 1 0 11 11 11 11 11 11 11 11 6 5 5 4:
04-01-198019:00 107 2 91 2 20 0 1 1 0 0 0 0 0 0 0 0 0 13 12 11 11 11 11 11 11 5 5 5 5:
04-01-198020:15 106 1 74 1 32 0 0 1 0 0 0 0 0 0 0 0 0 10 10 9 9 9 9 9 9 8 8 8 8:
```

C.   Data file BIERR.RAM, containing bushing insertion failure data.
     Similar to SIERR.RAM, for stud installation failures.
     (Random access file.)

FIGURE 2 - Message and data file organization.

ASYNC Interface

The robot sends a 4-byte hexadecimal word that the BASIC program converts into a 3-digit, base ten number.

The ASYNC protocol is as follows:

ROBOT:	"REQUEST TO SEND"
STATION CONTROLLER:	"OK TO SEND"
ROBOT:	"DATA"

Messages are received by the BASIC program and stored in a buffer whenever they are sent. The BASIC program sends the "OK TO SEND" message as soon as the "REQUEST TO SEND" message is retrieved from the buffer and decoded. If the "REQUEST TO SEND" is not answered within five seconds after it is sent, the robot sends it again. After an additional five second delay the robot continues with its program execution and "forgets" about sending the message. The "OK TO SEND" must be sent while the controller is expecting it. If it is sent when it is not expected an operator intervention may occur.

In order to avoid hard failures (a failure requiring an operator intervention) of this type the BASIC program must respond to incoming messages from the robot in less than five seconds in all cases.

Polling System

The BASIC program polls the input/output (I/O) lines in order to handle requests or incoming information. Figure 3 shows the general operation of the program. The time it takes for the program to poll all of the incoming lines and continue processing computations that are running (complete one "master cycle") must be kept to a minimum in order to maintain acceptable machine and operator response times and to enable the robot to operate properly.

Certain computations and transactions take more time than is allowed by the five second cycle time limit. In these cases execution is interrupted at a certain point, the master cycle of the BASIC program is performed, and the computation is resumed. Figure 4 shows the basic scheme for controlling sub-program execution that may span several master cycles.

The process of bouncing back and forth between sections of the program gives the illusion that several applications are being processed concurrently. While this is not the case, the desired effect is achieved.

The plotting of process control information requires the retrieval of large amounts of data from files on the hard disk. This is very slow and is only possible because of the ability to spread the total execution over several master cycles.

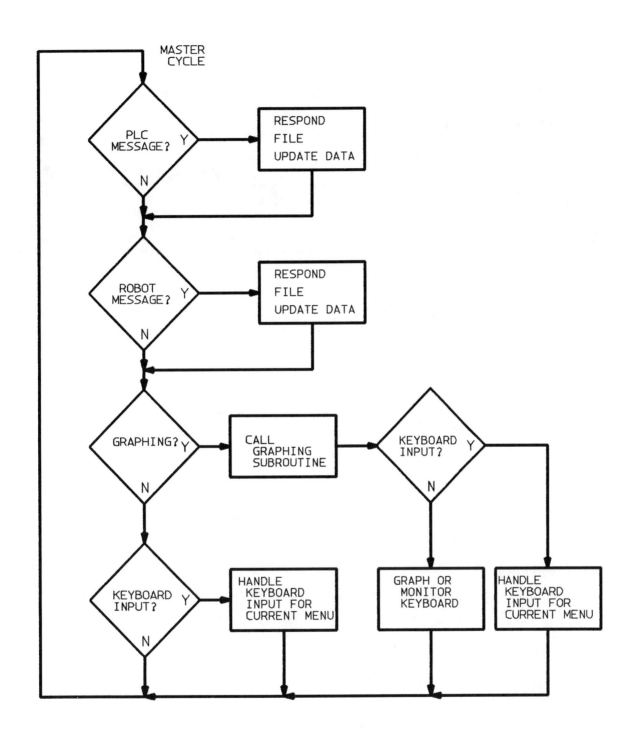

FIGURE 3 - BASIC CONTROL PROGRAM POLLING LOOP.

```
MAIN PROGRAM
      EXAMPLE = 0
      START
      .
      HANDLE ASYNC INPUT
      HANDLE PLC INPUT

      .
      IF EXAMPLE = 1 THEN CALL POINTER_EXAMPLE : GOTO END_OF_MONITOR
      .
      MONITOR KEYBOARD
          .
          .
          IF KEYBOARD_INPUT = N_{POINTER\_EXAMPLE} THEN CALL POINTER_EXAMPLE
          .
      END_OF_MONITOR
      GOTO START
END


---   ---   ---   ---   ---   ---   ---   ---   ---   ---   ---   ---   ---   ---


SUBROUTINE POINTER_EXAMPLE
          IF POINTER = 1 THEN GOTO 10
          IF POINTER = 2 THEN GOTO 20
          IF POINTER = 3 THEN GOTO 30
      POINTER = 0
      ----------------------------------------------------
      :               INITIALIZATION                     :
      : EXECUTE FIRST TIME SUBROUTINE CALLED             :
      ----------------------------------------------------
      EXAMPLE = 1
      POINTER = 1
      EXIT SUB
   10 ----------------------------------------------------
      : PROGRAM EXECUTION, PART 2                        :
      : EXECUTE SECOND TIME SUBROUTINE CALLED            :
      ----------------------------------------------------
      POINTER = 2
      EXIT SUB
   20 ----------------------------------------------------
      : PROGRAM EXECUTION, PART 3                        :
      : EXECUTE THIRD TIME SUBROUTINE CALLED             :
      ----------------------------------------------------
      POINTER = 3
      EXIT SUB
   30 MONITOR KEYBOARD
      IF KEYBOARD_INPUT = N_{HALT}
              THEN
                      POINTER = 0
                      EXAMPLE = 0
                      EXIT SUB
              ELSE
                      EXIT SUB
   END SUB
```

Figure 4 - Controlling polling frequency with pointers in BASIC.

STATISTICAL PROCESS CONTROL CHARTS

Code Structure

The graphing code is structured to complete its execution in eight or more cycles. That is, the pointer depicted in Figure 4 will take on eight values before the plot is completed and the program is monitoring the keyboard again.

The stages of drawing the plot are as follows:
1. Initialize the graphing parameters.
2. Get sample size from the operator.
   (Cycles until carriage return received.)
3. Get data from hard disk file. (May take multiple cycles.)
4. - 7. Print titles; draw data curve, moving averages and limits; label vertical axis.
8. Monitor keyboard.

By far the most time-consuming operation is getting the data from the files. The data are kept in random access files to minimize the access time. After the operator inputs the sample size, the BASIC program calculates the record number at which it must start in order to have the last machine produced included in the last data point plotted. In order for the BASIC program to check incoming ASYNC and PLC messages at no greater than 5 second intervals, data retrieval is suspended every 50 samples. This means that if the sample size is two, then all the data will be retrieved in one cycle, for a total of eight cycles to plot the data. If, however, the sample size is 200, then 80 cycles are required to retrieve the data (four cycles for each data point and 20 data points), for a total of 87 cycles to plot the data.

If the internal PLC is not operating and no operations are being performed, typical times to plot the curves are as follows:

CURVE	SAMPLE SIZE	PLOT TIME (SECONDS)
Daily Going Rate	20	11
	100	36
	200	44
Dual Screwdriver Failures	20	19
	100	75
	200	93
Operator Interventions	20	19
	100	76
	200	94
Bushing and Stud Failures	1	7
(groups of 100 attempts)	2	11
	5	23
	10	41

Data Storage

Process capability and graphing scale maximums are stored in individual data files. The values can be modified without re-compiling the code.

Daily going rate, dual screwdriver failures, and operator interventions are all stored in the same random access file, by machine. Bushing insertion and stud installation data are kept in separate files. The data are grouped in sets that represent approximately 100 attempts each. Therefore, if the operator selects a sample size of "1" for viewing bushing insertion attempts, he will be shown data representing the last 2000 attempts (20 points representing 100 attempts each).

The last data point displayed generally does not contain the full 100 attempts. It represents the failure rate data for the attempts that have been made since the last full group of 100 was completed.

Control Limits

The standard deviation calculations are appropriate for the type of data being analyzed.

Daily going rate and operator intervention time are treated as variables. The equation for the standard deviation is

$$\sigma = (\Sigma (Xm-Xmm)^2/n)^{1/2}$$

where,
        Xm = Each of the 20 sample means plotted
        Xmm = The mean of the plotted sample means
        n = Sample size.

Operator interventions are treated as defects per unit, with

$$\sigma = (Xmm/n)^{1/2}$$

where,
        Xmm = The mean of the plotted sample means
        n = Sample size.

The dual screwdriver, bushing insertion, and stud installation failures are treated as percent defective (p chart):

$$\sigma = (Pmm(1-Pmm)/n)^{1/2}$$

where,
        Pmm = The mean of the plotted sample means
        n = Sample size.

220

Graphing

One of the options on the main menu is for "Statistical Performance Graphing". Selection of this option causes the presentation of another menu (Refer to Figure 5, screen 1) that describes each of the general categories of performance data that can be plotted.

After a selection is made from the graphing menu the operator is queried for the desired sample size (Figure 5, screen 2). After the sample size has been input by the operator the data are taken from the appropriate data files, calculations are performed, and the process control chart is plotted. As the data are being acquired the monitor displays the status (Figure 5, screen 3). A period is printed each time all the data for one point on the curve is collected. The number of the data point being processed is printed each time the master cycle is executed. There are 20 data points plotted. Twenty-four data points are collected so that the five-group moving average is correct for the first data point.

The best way to describe the graphics display is to go through a typical plot in detail. Refer to Figure 5, screen 4.

SCALE:
The vertical scale is from 0 to 2000 boxes per shift. The minimum is fixed at 0 and the maximum is read from a disk file.

DATE STAMP:
The samples are dated across the top. The month, day, and hour that the last machine of the group was completed is above each point.

KEY:
The key in the upper left corner is a reminder that the point-to-point graph represents the averages of the individual samples, or groups; and, the step plot represents the five-group moving average. The first moving average plotted is based on the first plotted sample averaged with the prior four samples (which are not plotted).

SAMPLE SIZE:
The group size is shown to be 40 machines per data point.

CONTROL LIMITS:
The process average is specified in a data file and is represented by the solid horizontal line. The three, two, and one standard deviation limits are represented by the wide, medium, and narrow dashed horizontal lines, respectively. The points that indicate that the process is out of control are circled.

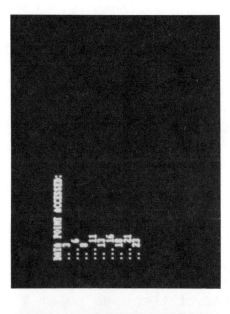

Screen 1
Main menu

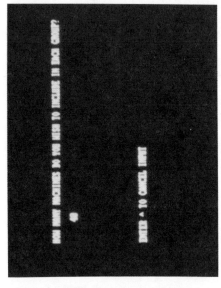

Screen 2
Sample size selection

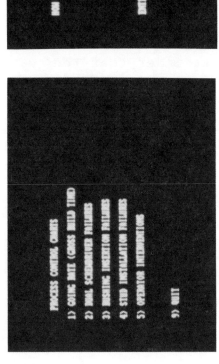

Screen 3
Data retrieval

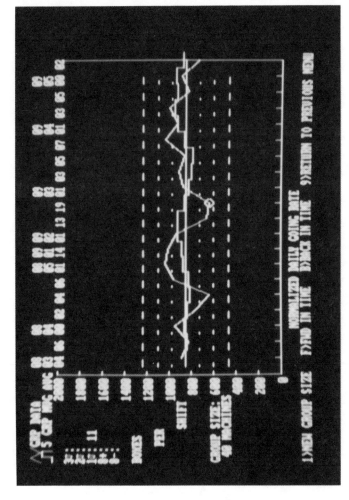

Screen 4 - Graphing

Figure 5 - Data retrieval
and graphing.
(Simulated data.)

The process is deemed to be out of control if:
- One point is more than three standard deviations from the
  process average.
- Two of three points are more that two standard deviations from
  the process average.
- Four of five points are more that one standard deviation from
  the process average.
- Eight consecutive points are on the same side of the process
  average.

The numbers below the KEY show which rule is violated by the
sample that is marked as out of control.

MENU:
The menu across the bottom allows the operator to choose a
new group size, go backward in time, or return to the process
control chart menu. This menu is different for each type of
graph.

CONTROL CHARTS

Figures 6 through 9 show each of the different type graphs
and how the different menu options are handled.

Daily Going Rate
(Figure 6)
Notice in screen B that the total data requested exceeds 20
data points. The five-group moving average then is not plotted
until the fifth data point is plotted.

Dual Screwdriver Failures
(Figure 7)
The two screwdrivers are identified as "near" and "far".
The holes they drive into are round and slotted. The part is not
oriented so each screwdriver drives into each type hole. The
robot controller is unable to detect either which screwdriver or
on which type of hole the tool failed. This information is input
by the operator when corrective action is taken. The circles in
screens A and B represent screwdriver re-tries. An increase in
the re-try rate precedes an increase in the failure rate.

If the operator selects "VIEW BY TYPE" from screen A, then
a cursor appears as in screen B. The cursor is positioned at the
left end of the range of data that are to be viewed by pressing
"L" or "R". Pressing "F" then fixes the left end of the cursor.
The right end is similarly moved to the right extreme of the
range of interest. Pressing "A" accepts the range and screen C
is produced. Screen C is drawn very quickly because the data are
already resident from the first acquisition.

Screen C displays the distribution of failures. In the
example shown there is no significant difference in the failure
rates by hole or by screwdriver.

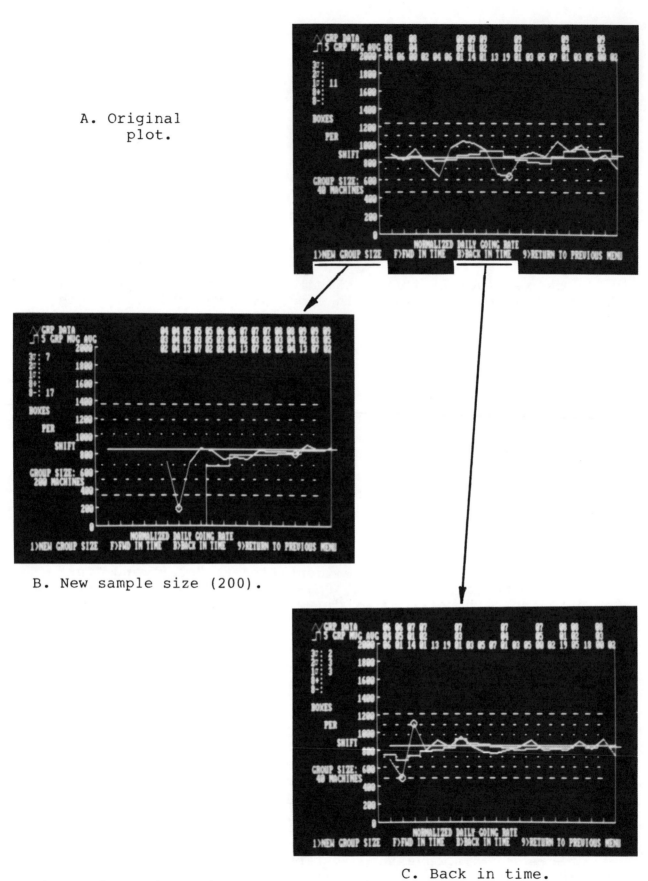

A. Original
plot.

B. New sample size (200).

C. Back in time.

Figure 6 - Daily going rate (simulated data).

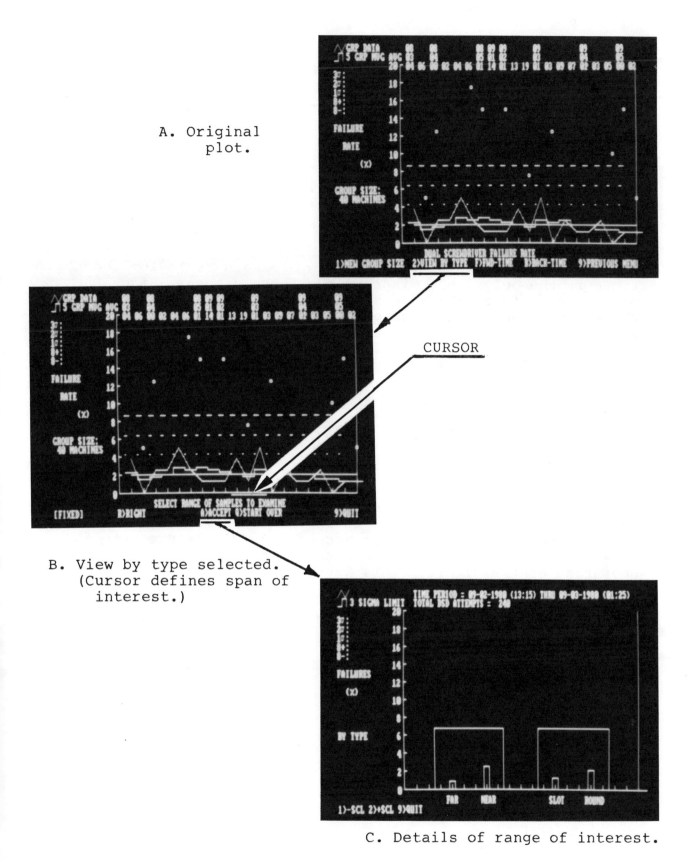

A. Original plot.

CURSOR

B. View by type selected.
(Cursor defines span of interest.)

C. Details of range of interest.

Figure 7 - Dual screwdriver failures (simulated data).

Operator Intervention
(Figure 8)

The duration of the operator interventions may also be displayed for the same time period as the rate graph.

Bushing Insertion Failures
(Figure 9)

The bushing inserter and stud installer both install their parts in each of 12 different holes: 8 holes in the "rear" cage and 4 holes in the "front" cage.

Screen A appears initially and shows the total failure rate for all bushing attempts. The three standard deviation control limit is the only limit displayed on these graphs. Since the failure rate information is summed into groups of approximately 100 attempts when it is stored, the sample size is not constant and the control limits are not constant. The data are not tested for out of control. It is left to inspection to determine which points are out of control.

The front cage failures are shown by selecting "2>FRONT CAGE" on screen A. This results in screen B. The failure rate of a particular front cage hole my be viewed by selecting the appropriate option from the screen B menu. In this case the failure rate of front cage hole 2 is shown in screen BB.

The failure rate over a specific time period may be viewed by selecting "3>BY HOLE" on screen A. The range of interest is selected by moving the cursor as explained previously. Screen CC is produced after the range is accepted. The failure rate for hole 2 of the front cage is seen to exceed the three standard deviation control limit. This is verified in screens B and BB.

The stud installation failure plots are very similar to the bushing insertion failure plots and are not included in the report.

SUMMARY

The system described allows on-line, real-time presentation of statistical process control charts without the need of an additional computer and without degrading the performance of the assembly station. The system is presently installed and operating on a manufacturing line in Austin, Texas.

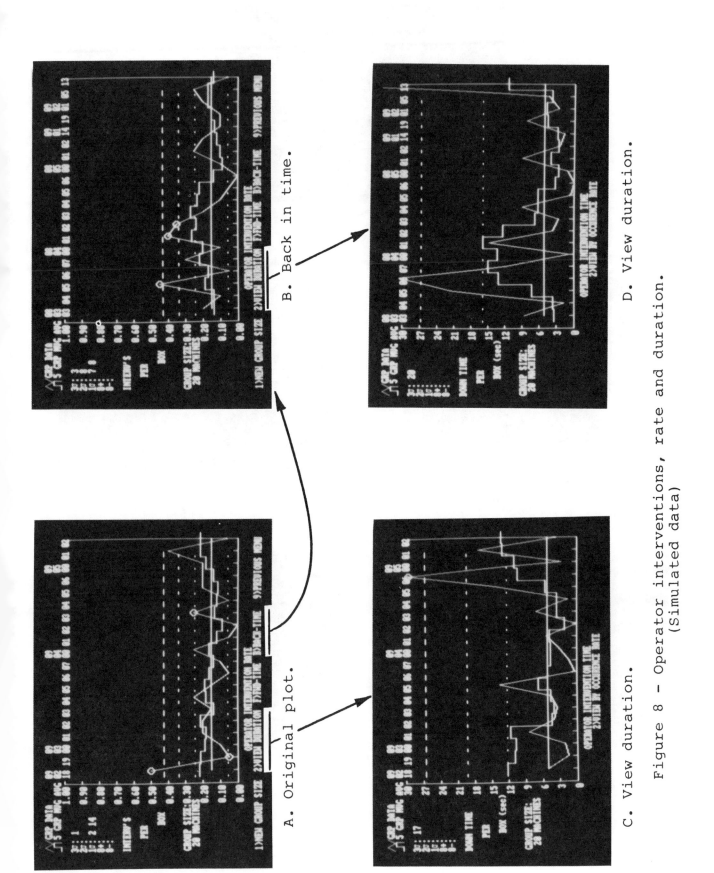

A. Original plot.

B. Back in time.

C. View duration.

D. View duration.

Figure 8 – Operator interventions, rate and duration.
(Simulated data)

227

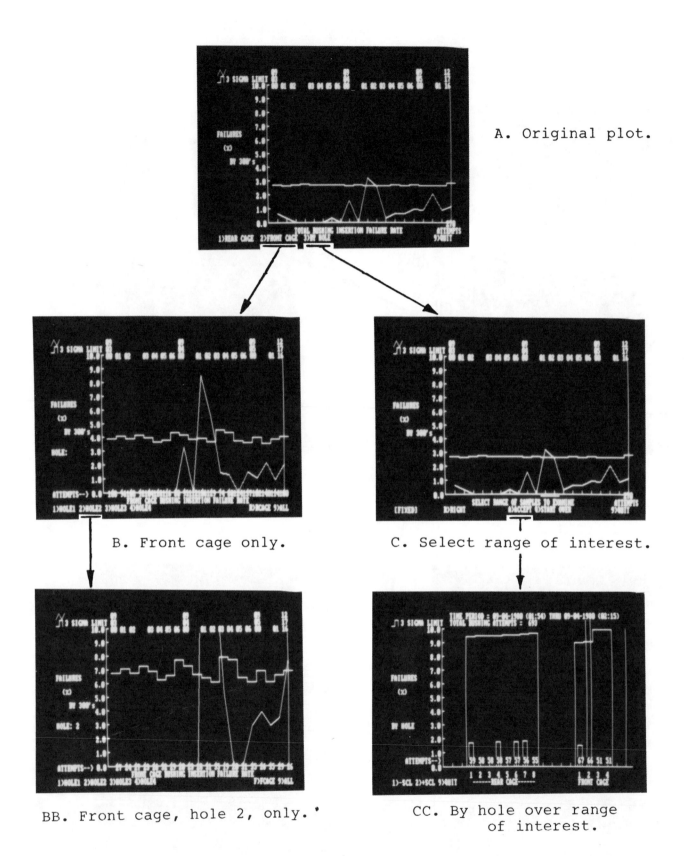

A. Original plot.

B. Front cage only.

C. Select range of interest.

BB. Front cage, hole 2, only.

CC. By hole over range of interest.

Figure 9 - Bushing installation failures. (Simulated data.)

Presented at the SME Synergy '86 Conference, June 1986

Statistical Process Quality Control
Can Be Used With Automated Assembly

Michael M. McRae
General Electric Company

The subject of this paper is self-explanatory and deals with the integration of statistical process quality control procedures with automated assembly equipment and some equipment and software which allow the process to be semi-automatic.

The attention that product quality has received in the last decade has been due to several factors. The major factors have been customer awareness and high quality foreign products. This desire to produce a better quality product has caused the manufacturers in this country to reexamine the quality assurance procedures developed in this country by individuals such as Messrs. Deming and Juran. These practices include Statistical Quality Control techniques.

The same foreign competition which has produced high quality products has offered, and continues to offer, products at a lower price because they have lower labor costs, a relatively high use of automation, and other modern techniques, such as Just In Time inventory systems.

To evaluate the use of Statistical Quality Control techniques in assembly auto-mation, it is appropriate to look at today's industrial climate in this country. There are several new technologies which have caused major changes in automated assembly equipment in the last ten years and will continue to modify the equipment and its function even more so in the future. These are:

- Robots are now starting to find their economic way into the assembly operations.

- PLC's were introduced just ten years ago and now PC's are interfacing with them on the shop floor.

- Vision systems are being used in inspection and in conjunction with robots.

- Lasers are being used for inspection, bar code readers, and for welding.

- CAD/CAM/CAE - The computer is truly now a manufacturing tool.

- CIM - The MAP, LAN networks are here and growing fast.

Another climatic condition which has caused major concern for American industry is the low productivity growth rate of recent years.

Many studies have shown a marked decline in productivity over the last twenty years in this country.

To add to the problems of our industrial climate, we continue to have increased labor cost. These costs consist of increased benefits and continuous wage increases. Even after recent labor concessions, our wages are still high when compared to other countries.

The loss of productivity and increased cost of wages have resulted in a loss of market share due to foreign competition. We can cry, "Foul!" or ask for protection by our government, but that is short-sighted. What needs to be done is to make changes which will allow American industry to be competitive, both in price and quality.

We must face the challenges of today and more that are to come. The most obvious of the possible solutions is to automate and integrate manufacturing processes. This will allow industry to take advantage of automation of product manufacturing and reduction of high waste and rework levels, as well as improved quality through improved utilization of inspection efforts.

Increased automation is a challenge because it requires investment and commitment. Increased automation is necessary to reduce the labor content of the product, but moreover a quality improvement accompanies automation of a product. Historically, the quality of automated assemblies has been more consistent than that of hand assembled-products. The reason for this is that automated equipment is usually a great inspector because, unlike humans, it cannot figure out how to use out-of-tolerance parts; it doesn't elect to overlook a detail that doesn't fit. Automated assembly equipment doesn't pry, bend, or hammer parts together. It simply defaults, jams, or stops.

In order to justify our automation expenditures, we must change our methods in determining saving to be derived by automating. The cost of poor quality in manufacturing in the United States is usually hidden in our operations and only recognized after it has been eliminated. That is part of the problem in justifying automation. By not identifying the cost of poor quality, we cannot quantify the improved return on investment of equipment and advanced technology. If we don't know the cost exists, we can't list its elimination as a savings. These costs include internal scrap costs, rework costs, warranty costs, high inventory costs, and lost repeat business. Thus, we have a higher selling price due to the aforementioned costs being built into pricing formulas--which of course adds to the difficulty in being competitive.

In order to improve the quality we must improve our utilization of the inspection that we presently do. Don't be confused. I am not saying that we must reject more parts because we plan to inspect to tighter limits, I am not saying we need to slow down the production cycle, I am not saying we must add more inspectors, and I am not advocating "inspecting in the QUALITY." What I am advocating is that we track the inspection that we do. I am advocating that we pay attention to the inspection already taking place in automated assembly equipment. Let me give you an example.

In performing its normal assembly process, an assembly machine performs inspections which may not have been planned but are inherent in its processes. An example is that when a feeder bowl aligns and feeds parts, it inspects out the bad units by dropping them back into the bowl or sometimes the part will jam in

231

the track. Usually, these jams are not recorded and therefore not tracked, but the automatic inspection is occurring.

As the parts are placed, riveted, welded, and assembled in various ways, more unintended but nevertheless automatic inspection takes place in the process. In addition, assembly machines usually are equipped with devices that check that the previous operation occurred properly and that the mechanism completed its stroke or operation. This information is available because each cylinder or check device has a separate I/O in the P.L.C. The PLC commands the mechanism to work and receives back a signal indicating that the cycle is complete, or not completed appropriately.

The C.P.U. of most assembly equipment available today, has all the necessary functions which would allow storage of each operation's signal as the equipment operates. But in order to record each and every failure, i.e., FAULT, one would need some major memory expansion and/or some added interfacing to other computers. Therefore, it is possible that we can use this information to generate statistical data if we could use it to our advantage.

The major item needed to track the faults statistically is more computer memory to store the extensive data. Exactly how much additional memory will be in direct relationship to the amount of I/O to be monitored and the amount of time that the stored information is to be retained/collected. As an example, let's take a simple case. If we look at a single station of a normal fixed automation assembly machine which feeds and places a part, the following I/O for continuing the individual mechanisms on the station would be:

OUTPUTS:

1. Feeder bowl power on/off.
2. Station stop (to cylinder valve).
3. Station release (to cylinder valve).
4. Shot pin engage (to cylinder valve).
5. Shot pin disengage (to cylinder valve).
6. Pick and place vertical up (to cylinder valve).

7. Pick and place vertical down (to cylinder valve).
8. Pick and place horizontal out (to cylinder valve).
9. Pick and place horizontal in (to cylinder valve).
10. Pick and place jaws close (to cylinder valve).
11. Pick and place jaws open (to cylinder valve).
12. Linear feed track power on/off.

INPUTS:

1. Feeder track low level electric eye.
2. Feeder track high level electric eye.
3. Sense carrier coming switch.
4. Carrier in place switch.
5. Station stop extended switch.
6. Station stop retracted switch.
7. Shot pin engaged switch.
8. Shot pin disengaged switch.
9. Pick and place vertical up switch.
10. Pick and place vertical down switch.
11. Pick and place horizontal out switch.
12. Pick and place horizontal in switch.
13. Part in gripper electric eye.

The input signals can be compared to an expected result or used to cause a fault if they do not occur in the proper sequence. Suppose the station cycle time was 6 seconds (which is typical for a nonsynchronous power and free machine). That would mean that it is possible that we would want to store either the fault only or the actual signal at a rate of 10/minute. Depending on our sample rate and sample interval, there is the possibility that we might want to store as many as 7800 signals per hour (10 X 13 X 60). If we collect the information for an 8 hour shift, that would be 62,400 signals. This is just for one simple station that does not have dynamic measurement capability. But why store all those signals? That's a good question. We probably wouldn't want to have all that information initially, but after we have had the machine for a year or more and as we learn about the station we may want to monitor the linear vibratory feed track level signals versus the on/off time required of the feeder bowls versus station cycle time, or the pick and place actuations in the vertical mode, etc. The importance of keeping track of the data on the station is that the signals allow us to be able to analyze the machine over time. This will allow us to determine patterns for improvement of the reliability of the equipment, as well as areas which need more frequent maintenance or even design modification. This all relates to cost of operation and product consistence.

One of the best methods of analyzing the data is through the use of Statistical Process Quality Control. One of the major advantages that we have available to us today which did not exist 30 years ago, when SPQC was originated, is that the data can be automatically collected and generated. I will discuss that process later.

S.P.Q.C. is a procedure of monitoring to evaluate and analyze the quality of a process. When the process is controlled accurately, its quality can be improved within the limits of the process capabilities and, as I indicated earlier, if the quality or reliability desired is above the process capability, then a design change can be made to improve the process. Therefore, by using statistical methods and measures of a process as it is at work, the quality can be measured and predictions made as to the continuation of the quality of its output and even predict, with a high degree of accuracy, the future level of the output quality.

The most commonly used statistical formulas and information are the chart types--the most common of which are:

- $\overline{X}$ Chart = average variables data - measurement

- R Chart = range variables data - measurement

- P Chart = fractional defects attribute data - yes/no

- C Chart = number of defects attribute data - yes/no

- Sigma chart = standard deviation

All of these charts appear to be similar. They all have upper and lower control limits which are calculated using the standard formulas found in any Quality Control Handbook. The control limits can and should be adjusted with experience of the process being monitored.

The most common of these is the $\overline{X}$ Chart. The points of the chart are determined by taking a sample of a population of readings and determining the

$\overline{X}$ (average or mean) value. The vertical coordinate of the chart is the measurement axis and the horizontal is the time axis.

For example, if you took a sample of 25 readings and calculated the average of those readings, and the value was .70 it would be the first point on a chart. Then later, if you were to take another sample of 25 pieces and determine that $\overline{X}$ value to be 0.77, it would be used as the second point, and so on.

The Range Chart appears similar to the $\overline{X}$ Chart; the difference is that the points on the chart, instead of being averages of sample readings, are the value of the range of the sample data. So when a sample is taken, the average is determined and saved for charting on the $\overline{X}$ Chart, and the range of the sample values is calculated and saved for charting on the R Chart. Imagine, if you will, a list of readings of measurements which are derived from a sample of production parts-- the average of those readings is the first point on the $\overline{X}$ Chart and the total spread or range of those readings is the first point on the R Chart. After obtaining many points from repeated sampling over time, we can plot the points and analyze the two charts. Thus, it can be determined if the process is stable and/or headed out of control.

These charts are for quantitative data or variables data, such as a measurement of pressure, distance, dimensions, torque values, etc. On an assembly we might monitor a voltage or pressure of a specific operation to obtain variables data. For those signals or pieces of information which are yes/no or go/no-go type information, a different set of charts is used. They are the P Chart and C Chart. These are attribute data charts.

The P Chart or proportion defective is similar to an $\overline{X}$ chart since it is a calculated value like the average for $\overline{X}$ Chart, but the P Chart uses the number of defectives in the sample divided by the number of items in the sample, or an average of the defects. Once again, this is an average.

Returning to a single station of a machine, the important information presently being wasted is that the station did not perform properly. If all the signals from this station are stored for later use, they can become control chart information on a P chart and can also be used for a C Chart.

Having all the data stored over a rolling 8-hour period allows any type of analysis to be performed. If we had a P chart on the electric eye which monitors the part presence in the gripper jaws, we could determine a trend that slowly the gripper was placing less and less (while not excessive, relative to a defect rate). This could signal that the plating or even part dimensions were changing and a correction could be made before--let me stress BEFORE--the rate of missed placements showed up as a problem under normal conditions. Then method, then, is really for defect prevention.

So now that we have generated the charts, what do we do with them? Having the data and not using it is like watching a pot boil over and not doing anything about it--interesting to watch, but watching alone does not correct the situation.

Now the action part comes in--we must find out why the station is not placing the part. It could be that the mechanism needs adjustment, or any number of other problems.

But who in the organization should receive this data first? If the President of the company is given the data first, all he/she would do with it is to tell someone else to do something with it or about the situation, and likewise the managers, and quite honestly, often so would the Q.A. department. All too often, if the information is fed directly to a mainframe computer in a CIM system, the Q.A. department analyzes the data and goes to a meeting where the information is a surprise to the production department. Although the Q.A. department should get the information, the person who should have the information first is the person responsible for the process, i.e., the operator or foreman over the area.

The operator or foreman should use the information to find the problem in time to improve the quality of the product and certainly prior to the problem causing the equipment to be shut down. Since the Q.A. department would also be able to monitor the data that the foreman has at his/her disposal, either on the spot or through a CIM system, they could see just which department/equipment is maintaining the best product quality level.

Assuming that we are convinced that we need S.P.Q.C. generated by the equipment, how and where do we get this system and information?

Now we have come to the real meat.

Unfortunately, it isn't all that simple to obtain. There are many ad or equipment and software which provide "data reporting systems." There are many new interface equipments available. Unfortunately, in many cases, both the software and hardware claims are largely exaggerated. It is difficult to separate the claims from the reality. Therefore, in order to do so it is necessary to understand how a typical system operates--which, hopefully, will help us ask the right questions of those who supply these items.

The basic elements of the system that is now becoming available are the equipment PLC, an interface box, a PC, a printer/plotter, and software. (REFER TO FIGURE 1) A system is needed that will store the signals both quantitative and/or qualitative from the measurements and the yes/no inputs and outputs of the PLC--then after obtaining this information on a real time basis it must be stored in memory. Once stored, it should be statistically analyzed periodically (sampled) and the results put into a Q.C. format.

(REFER TO FIGURE 2) There are several software packages available.

There is one system that presently is advertised to have the interface hardware and software as an integral package. The keyboard, CRT, and printer/plotter must be customer supplied. This is the Gidding & Lewis PIC (Programable Industrial Computer) with the "STAT-PAC" software. Handling only variables or measurement type data, the STAT-PAC capability includes subgroup range sizes up to twenty (20), for a data pool of up to five hundred (500) values per station. In the real time data mode, data is collected from each station and stored in the control's memory. A total of 500 points per station can be kept. In the display mode, graphs can be displayed on the CRT. The data can also be transmitted to printers for permanent records. STAT-PAC has an automatic roll-over feature which adds new data and drops the oldest previous data automatically. The average selling price of G&L PIC interface and software is just under $8,000. This price does not include a printer, monitor, or keyboard.

Allen Bradley Company has an interfacing industrial hardware which can be used with many of their PLCs. The software which they recommend for the S.P.Q.C. function and others is "FACTORY CALC" from Misa; however, I have no other verifying information on FACTORY CALC or Misa Company.

Texas Instruments has an interface and display terminal, model CVU 5000, which can be used with many of their PLCs as well as Allen Bradley and Modicon equipment. It can provide management data collection and storage and display information. So far, it is not advertised to have SPQC information. TI also has a network controller model ICC 6000 which is handled by the "TIWAY" network system. It could be used for QC historical information, but it is a CIM type system.

One other piece of equipment that I am aware of is a hand-held PC, specially designed and programmed by Sharp Corporation, which is available from Elco Industries. It provides variables and attribute data and charts, but at present it does not have any interfacing capabilities.

I would like to point out that there are many special systems in existence today which accomplish the functions I am describing here, but these are usually special designed and built for that application. What I am proposing is a more standard off-the-shelf type system which can be used with standard PLC's and PC interfaces.

On the positive side of the picture, I do have some information on sources of software packages.

Several software packages that run statistical quality control programs are available for use with personal computers. Starting at that point in the system, we can work backwards to achieve our desired result.

P.Q. Systems has a package called SQC PACK. SQC PACK is very complete in its Q.C. information. Its capabilities include VARIABLES DATA (X and R, Sigma, median information); ATTRIBUTE DATA (P, NP, C, U information) and PROBLEM SOLVING ("PARETO" and Cause & Effect information). The software is menu driven for operator ease of use. It is compatible with IBM PC's, PC/XT's, PC/AT's, which have 192K memory, two (2) disk drives, and a graphics plotter. Also, the software can be used on the IBM 5531, 7531 and 7532 industrial computers accompanied by IBM 5532 or 7534 displays and/or 5533 graphics printer and cable. The package also has been adapted to IBM compatibles such as NCR PC4, COMPAQ, EAGLE, TANDY 2000, ZENITH and others. P.Q. Systems also offer a more basic package SQC PACK for APPLE, T.I., and VICTOR 9000 computers.

Many popular printers and plotters can be supported by the SQC PACK software package. This package can obtain input from popular spreadsheets such as LOTUS 1-2-3 and DATABASE programs such as DBASE II & III, among others. The control charts in this presentation were furnished by courtesy of P.Q. Systems and are typical of their CRT screen format. SQC PACK is available directly from P.Q. Systems and is also distributed by IBM Industrial Computer Systems Group. The price is less than $600.

A software package priced at less than $800 is QUALITY ALERT from Penton Software, Inc., also available from IBM Industrial Computer Systems Group. QUALITY ALERT is a menu-driven package whose functions include data storage and manipulation of disk data files. The statistical quality control (SQC) functions include:

- ° descriptive statistics

- ° histograms

- ° control charts

- ° process capability

- ° generation of variable & attribute data samples

QUALITY ALERT will run on IBM 5531 Industrial Computer Model 2 with 256K memory, IBM 5532 Display and IBM 5533 Graphic Printer. Professional Applications Development has just released "STATPAD" which is a menu-driven, LOTUS 1-2-3 compatible Q.C. program. It sells for $65, although I cannot comment on its capabilities.

While these software packages run on PCs and industrial computers, added software is required to obtain the data directly from the assembly machine PLCs. One of these driver software packages is available from Soft Systems Engineering, Inc. SSE has several packages available which interface with popular PLC's such as Allen Bradley, T.I., and Modicon.

One package called TAXI is an application software program for moving BCD data from A/B PLC-2 family PLCs to LOTUS and VISICALC spreadsheets. Remembering that SQC PACK from P. Q. Systems can access LOTUS, it would appear that if you had an A/B PLC 2, an IBM P.C., a printer/plotter and SQC PACK, SSE TAXI, and LOTUS software, you could access any desired I/O or variable data area of the assembly machine PLC and generate any or all of the statistical quality information you desire.

This set--a prototype of a standard off-the-shelf type system--will be verified by Benerson Corporation in the near future as we are working on this method for a present customer. It may be possible that some special programs for interfacing will be required. One area in question in this process is: Will the system run the SQC information while collecting data in real time, or will a second PC or a special co-processor board be required?

As a case in point, at Benerson Corporation we were able to sell two customers on this very concept. Then it came time to implement the process. Both applications have not come to pass--the first because of the sheer size of the job, which is not complete as of this date--and the second because the customer could not decide what he wants to monitor and how often and what the data accumulation time should be. What we had planned to do was to use the "SSE TAXI" package--the "SPQ PACK" as described just now--but it didn't work. The reason was that "SPQ PACK" could not be interfaced to the TAXI software. The end result was to use the TAXI software to address the CPU in the machine to obtain the digital information and then to generate our own software for the statistical data desired. This makes the system less flexible than was originally desired, because it became a tailor-made package versus a generic package.

Assuming that you locate a source for a system (I recommend a full system integrator who understands both equipment and software), the next most important item to consider is training.

Assuming the S.Q.C. data is being generated automatically, one needs to know how to analyze the information to understand the cause of the problem that the data is pointing to. Then corrective action can be taken. Said another way, this is simply understanding the application of S.P.Q.C. For S.P.Q.C to be truly effective in correcting problems, the operator/foreman has to be educated in what S.P.Q.C. information is, why it is important, and how to generate it and use it.

This means also that the operator must understand the system capabilities. When I say system in this case, I am defining system as the assembly equipment and the interface S.P.Q.C. software, terminal, and printer/plotter. Now we have broadened the operator/foreman's area to include the normal understanding of the equipment and the S.Q.C. processes. This is the area in which added training is usually required. This additional training requires classroom courses, manuals, etc.

An understanding of the system capabilities will lead to understanding the limits of the system. Truly knowing the equipment means one understands that the equipment cannot be expected to make good assemblies out of bad parts, nor can the S.Q.C. information say how to adjust the machine. Therefore, the operator must know the system in order to correct those things which are causing the problems—this means analyzing the data and adjusting the equipment.

Another aspect of the system which must be understood is how to adjust the actual UPPER and LOWER CONTROL LIMITS STATISTICALLY which alert us that our equipment is truly out of control.

Who provides training in methods of statistical quality control? Most training thus far in the U. S. has been in-house training courses; however, a training course is available from Control Data Corporation. This is instruction in Statistical Quality Control not only for people on the production floor but for Q.A. personnel as well. It teaches awareness of the total quality system, or what is sometimes called Total Quality Control.

Training of our people is the key and, although listed last, is the most important element of the system.

In summary, then, the elements for our future are:

1. We need to meet the foreign competition head-on and in many ways.

2. One way is better product quality in our automated assembly process.

3. By adding software and computers to our assembly equipment, the operator can monitor the quality of the output using SQC techniques.

4. In order to do this, we need to train our operators in SQC.

The bottom line that we are all after, then, is a BETTER PRODUCT with HIGHER QUALITY/RELIABILITY at a CHEAPER PRICE. And the reason is: we want to regain control of our country's manufacturing spot in tomorrow. We want to secure our industry's future in the marketplace of tomorrow.

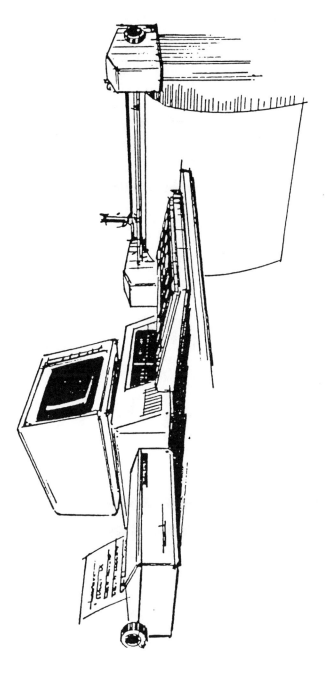

FIGURE 1

243

SOFTWARE PACKAGE NAME	COMPANY	VARIABLES DATA	ATTRIBUTE DATA	COMPARABILITY	INTERFACE DIRECTLY TO PLC	INTERFACE DIRECTLY TO PC	AREA PRICE
STAT-PAC	GIDDING & LEWIS	YES	NO	GIDDING & LEWIS PIC ONLY	YES PIC ONLY	NO	$8,000
FACTORY-CALC	MISA/	YES	YES	ALLEN BRADLEY	YES A.B. ONLY	NO	
SQC PACK	P.Q. SYSTEMS	YES	YES	IBM, NCR, COMPAQ EAGLE, TANDY 2000, ZENITH, APPLE, TI	NO	YES	$1,000
QUALITY ALERT	PENTON SOFTWARE	YES	YES	IBM	NO	YES	$1,000
TAXI	SOFT SYSTEMS ENGINEERING	NO	NO	MOST PLCs TO PCs B.&D. FROM PLC TO PC TO LOTUS OR VISICALC	YES	YES	$1,000

FIGURE 2

Presented at the SME Synergy '86 Conference, June 1986

Statistical Process Control in Group Technology Production Environment

Dr. Johnson Aimie Edosomwan
IBM Data Systems

INTRODUCTION

In this era of technological explosion, competitive environment demand a better quality product or service at the existing price or at a lower price. As a result, quality has emerged as a major new business strategy in many organizations. The strong emphasis on quality is driven by a number of reasons, including:

1. Acceptance of products produced or services rendered relies on conformance to requirement or specification, which means all output produced must be defect free.

2. Quality improvement increases productivity.

3. Increasing cost of production or services (labor, materials, energy, capital, and technology).

4. Product liability could lead to loss in market share.

5. Increasing consumer education and awareness of quality.

6. Global intensive competition in all industrial sectors.

As shown in Figure 1, poor quality goods and services can have significant impact on the survival of any economic unit. In order for organizations to effectively offer good quality products and services that will compete in the world market, it is necessary for them to institute a statistical process control system. Such a system can have the following benefits:

1. Statistical process control facilitates process capability improvement. It helps in determining problem root cause and monitoring of corrective actions.

2. Statistical process control facilitates increases in productivity through control of rework machine down time, scrap, and work-in-process inventory.

3. Statistical process control provides a basis for both management and employee to understand and control assignable causes that affects a process performance.

4. Statistical process control facilitates the minimization of the "cost of quality". The expense of doing things right the first time.

5. Statistical process control creates a basis for effective control and understanding of the complex interaction among production and human related variables such as: Machines, fixtures, materials, and human work habits and efficiency.

6. Statistical process control facilitates improvements in quality yield and reduction in product cycle time.

7. Statistical process control tools such as control charts provide a common language for communications about the performance of a process.

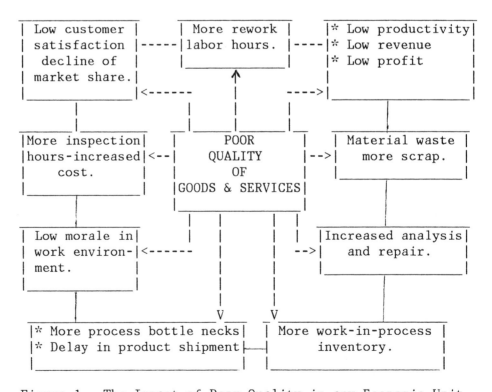

Figure 1. The Impact of Poor Quality in any Economic Unit

DEFINITION OF A PROCESS CONTROL SYSTEM

A process control system is a feedback mechanism that provides information about the process characteristics and variables, process performance, action on the process inputs, transformation process, and action on the output. The major components of a process control system are shown in Figure 2.

THE KEY REQUIREMENTS OF PROCESS MANAGEMENT

In order to improve the quality of products or services from a process, the following basic requirements are necessary:

1. Management philosophy must be one that is geared towards continuous quality improvement at the source of production or service. Process control is not a one time action. It is a continuous feedback mechanism that involves both management and employees, in producing good products and services.

2. There has to be a system for measurement, control, evaluation, planning and improvement.

3. There has to be a data base for the storage, retrieval, and access of information about the process.

4. There has to be a clear definition of process parameters and variables. Input and output elements must be clearly identified.

5. All process parameters have to be characterized with complete understanding of repeatability and variability.

6. Process ownership by everyone involve in the management process.

7. Management philosophy has to be that no level of defect is acceptable.

8. Quality improvements from key process control action must be documented and rewarded on a timely basis.

9. The management information system for process control must provide a real time feedback mechanism to everyone involved in the quality improvement effort.

10. Provision for on-going training on process control techniques.

11. The attainment of good quality product should be through prevention of defects and not through detection of defects.

12. The removal of defects from a process should be done through root cause analysis and the implementation of proper corrective action. The costly inspections should be discouraged. The strategy should utilize sampling inspection where necessary and no inspection when not warranted.

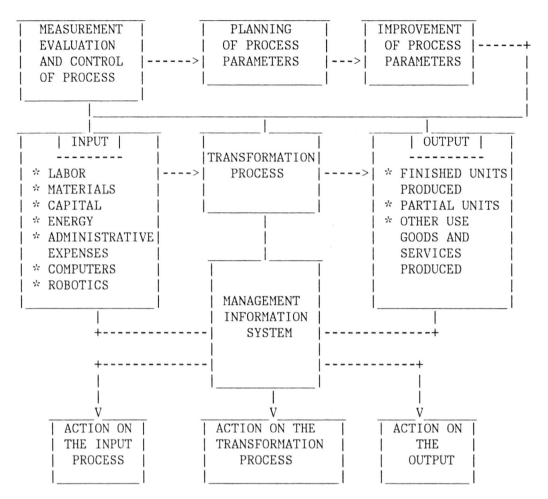

FIGURE 2. COMPONENTS OF A PROCESS CONTROL SYSTEM.

DEFINITION OF GROUP TECHNOLOGY

Burbidge [1] define Group Technology (GT) as a technique for production management which seeks to obtain similar economic savings in batch and jobbing production, to those already achieved using line flow in the simpler process industries, and in mass production, and second to provide a better type of social system for industry, in which improved labor relations are easier to achieve. For the purpose of this paper the above definition of group technology will be adopted.

IMPLEMENTATION METHODOLOGY FOR STATISTICAL PROCESS CONTROL IN GROUP TECHNOLOGY PRODUCTION ENVIRONMENT

A ten-step approach shown in Figure 3 should be followed when implementing statistical process control in a group technology production environment. The steps shown in Figure 3 are offered as a guide and could be followed with minor modifications depending on the type of production environment. However each element within each step is important to the successful implementation of statistical process control (SPC) in group technology environment. Each step in the implementation methodology will now be described.

248

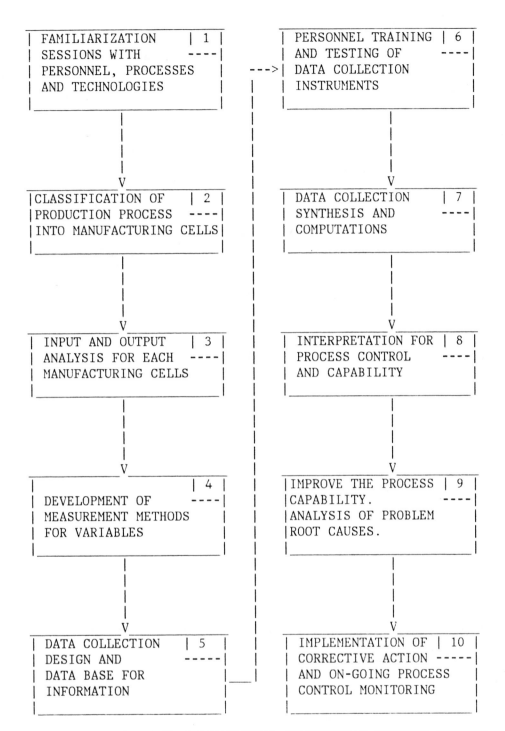

```
| FAMILIARIZATION    | 1 |          | PERSONNEL TRAINING | 6 |
| SESSIONS WITH      ----|          | AND TESTING OF     ----|
| PERSONNEL, PROCESSES |     --->| DATA COLLECTION    |
| AND TECHNOLOGIES   |     |    | INSTRUMENTS        |
|_____|     |    |_____|
         |                 |             |
         |                 |             |
         |                 |             |
         V                 |             V
|CLASSIFICATION OF   | 2 |  |    | DATA COLLECTION    | 7 |
|PRODUCTION PROCESS  ----|  |    | SYNTHESIS AND      ----|
|INTO MANUFACTURING CELLS| |    | COMPUTATIONS       |
|_____|     |    |_____|
         |                 |             |
         |                 |             |
         |                 |             |
         V                 |             V
| INPUT AND OUTPUT   | 3 |  |    | INTERPRETATION FOR | 8 |
| ANALYSIS FOR EACH  ----|  |    | PROCESS CONTROL    ----|
| MANUFACTURING CELLS |    |    | AND CAPABILITY     |
|_____|     |    |_____|
         |                 |             |
         |                 |             |
         |                 |             |
         V                 |             V
|                    | 4 |  |    |IMPROVE THE PROCESS | 9 |
| DEVELOPMENT OF     ----|  |    |CAPABILITY.         ----|
| MEASUREMENT METHODS |    |    |ANALYSIS OF PROBLEM |
| FOR VARIABLES      |     |    |ROOT CAUSES.        |
|_____|     |    |_____|
         |                 |             |
         |                 |             |
         |                 |             |
         V                 |             V
| DATA COLLECTION    | 5 |  |    | IMPLEMENTATION OF  | 10 |
| DESIGN AND         -----|  |    | CORRECTIVE ACTION -----|
| DATA BASE FOR      |___|    | AND ON-GOING PROCESS |
| INFORMATION        |          | CONTROL MONITORING  |
|_____|          |_____|
```

FIGURE 3: STEPS FOR IMPLEMENTING STATISTICAL PROCESS CONTROL
IN GROUP TECHNOLOGY PRODUCTION ENVIRONMENT.

STEP 1: FAMILIARIZATION SESSIONS WITH PERSONNEL, PROCESSES, AND
TECHNOLOGIES

In this step a formal study and understanding of the production
processes, procedure, cost accounting information, existing
data collection system, types of products produced with
associated technologies. A meeting should be held to introduce
purpose of the statistical process control in the work environment.

STEP 2: CLASSIFICATION OF PRODUCTION PROCESS INTO MANUFACTURING CELLS.

The production processes should be divided into manufacturing cells. The cells should have the property of being able to process a family of products that share at least one common property, either design or manufacturing or both. Other factors to consider are workflow pattern, source of delivery of raw materials, ability to obtain a balanced manufacturing process - using line balancing approach, and interdependencies among production variables and manufacturing cells.

STEP 3: INPUT AND OUTPUT ANALYSIS FOR EACH MANUFACTURING CELL

In order to understand the characteristics of all production variables, input and output analysis of each cell is performed. The significance of each task performed, and the characterization of both equipments and processes are obtained. For example equipment characterization should consider items such as: variables that affect machine output, machine capability (uptime and downtime of machine), output per hour, repeatability of machine problems, etc. Process characterization should consider items such as incoming quality of products, the impact of multiple products on one machine, process yields, man-machine interface problems. The emphasis should be to determine the variables that affect process output and their effects on each manufacturing cell and the total combination of cells. This will enable us to understand how each variable in each cell can be controlled.

STEP 4: DEVELOP MEASUREMENT METHOD FOR VARIABLES

The primary goal should be emphasis on the use of control charts in the understanding and reduction of fluctuations in a process until they are in a state of statistical control at the level desired. By definition, a control chart is statistical device used for analysis and control of repetitive processes. They are used to study variations in a process that are attributed to special causes. An example of measurement methods for attribute control charts is summarized in Table 1.

STEP 5: DATA COLLECTION DESIGN AND DATA BASE FOR SPC INFORMATION

In this step various forms and instruments are design to capture input and output components needed for statistical process control in each manufacturing cell. In designing these forms various factors such as: frequency of data collection, information required, computation steps, scale of measurement and clarity should be considered. An example of measurement form for variable control chart calculation work sheet is shown in Table 2. A data base should be set up for the storage and retrieval of information on process performance at each manufacturing cell. Personal computers, Series I computers and others are useful systems for data base management.

TABLE 1: <u>MEASUREMENT METHOD FOR ATTRIBUTES CONTROL CHARTS</u>

What is measured	Control Chart	Sample Size	What is to be controlled	Center Line	Control Limits	Comments
Number of defectives in sample	np Chart	Constant	d: The number of defectives in a constant sample size	$\bar{np}$	$\bar{np} \pm 3\sqrt{n\bar{P}(1-\bar{P})}$	n = Sample size $p = \dfrac{\text{Number defectives}}{n}$ $\bar{p} = \dfrac{\text{Total defectives}}{\text{Total inspected}}$
Average number of defectives in sample	p Chart	Varies	P: The ratio of defectives to sample size	$\bar{p}$	$\bar{p} \pm \sqrt{\dfrac{\bar{P}(1-\bar{P})}{n}}$	
Number of defects in sample	c Chart	Constant	c: The number of defects in a constant sample size	$\bar{c}$	$\bar{c} \pm 3\sqrt{\bar{c}}$	$c = \dfrac{\text{Total \# defects}}{\text{Number of samples}}$ More than one defect be recorded on a piece in the sample
Average number of defects in sample	u Chart	Varies	u: The ratio of nonconformities to sample size $u = \dfrac{c}{n}$	$\bar{u}$	$\bar{u} \pm 3\sqrt{\dfrac{\bar{u}}{n}}$	c = # of defects $u = \dfrac{\text{\# defects}}{\text{Sample size}}$ $\bar{u} = \dfrac{\text{Total \# defects}}{\text{Total pieces inspected}}$

TABLE 2:<u>CALCULATION WORKSHEET FOR VARIABLE CONTROL CHART</u>

```
==============================================================
|                 CONTROL LIMITS based on subgroups          |
|------------------------------------------------------------|
|    Calculate Average Range:                                |
|           i                                                |
|        ___ R                                               |
|        \                                                   |
|    R = ---- = ---------- = ----------                      |
|         k                                                  |
|    ___                                                     |
|    \  = Sum of, and k = number of subgroups                |
|    /                                                       |
|                                                            |
|                                                            |
|    Calculate Control Limits for Ranges:                    |
|                   _                                        |
|    UCL   = D X R = _____  X _____ = _____       |
|       R    4                                               |
|                                                            |
|                   _                                        |
|    LCL   = D X R = _____  X _____ = _____       |
|       R    3                                               |
|_____|
|                                                            |
|    Calculate Grand Average:                                |
|                _                                           |
|    =      ___ X                                            |
|    X = ---- = ---------- = ----------                      |
|           k                                                |
|                                                            |
|    Calculate Control Limits for Ranges:                    |
|        _                                                   |
|    A R   = _____  X _____ = _____              |
|     2                                                      |
|                =                                           |
|    UCL   = X + A R = _____  + _____ = _____       |
|       X        2                                           |
|                                                            |
|                =                                           |
|    LCL   = X - A R = _____  - _____ = _____       |
|       X        2                                           |
|_____|
|                                                            |
|    Estimate of Standard Deviation (If the process is in    |
|    statistical control):                                   |
|                                                            |
|    ^       R                                               |
|    o = --- = _____                                     |
|         d                                                  |
|         2                                                  |
==============================================================
```

STEP 6 PERSONNEL TRAINING AND TESTING OF DATA COLLECTION
        INSTRUMENTS

In this step, personnel associated with the implementation
of statistical process control in each manufacturing cell
should be trained. Example of a typical training agenda

is shown below:

(1) Rationale for manufacturing cells and concepts of group technology

(2) Statistical process control concepts and key definitions.

(3) Construction and interpretation of statistical control charts.

(4) Methodology for process analysis, the role of the engineer in the process improvement.

(5) Experiment design methods.

(6) Variable mapping.

(7) Dependent and independent variable identification.

(8) Data collection and analysis format.

(9) Statistical tools for data analysis.

(10) Group case studies on high flier defects.

STEP 7: DATA COLLECTION, SYNTHESIS AND COMPUTATIONS

Using the various forms designed in Step 5, input and output components, process yields and measures for each manufacturing cell are collected periodically. The synthesized data are then used to compute control limits and process averages for each manufacturing cell.

STEP 8: INTERPRETATION FOR PROCESS CONTROL AND CAPABILITY

The process yields (defects, UCL, LCL, process averages, etc.) of each manufacturing cell are analyzed periodically seeking why they increased or decreased. The emphasis should be placed on understanding the assignable causes for variation with the process. Points outside control limits shows a special cause that should be investigated urgently. Also a trend might indicate process drift. Six or more consecutive points indicated that the process is drifting. The introduction of a new parameter or tool to the process can lead to a shift in the process. For example, at installation of a new machine. It is important to watch out for cycles of consistent repeated problems within the process. Cycle may be caused by variations in process parameters such as speed, temperature, operator skill levels, and product mix. Instability factors are normally caused by improper tuning of process parameters such as maladjustment in machines. The most difficult points to understand in control charts are freaks. These are points that lie outside the control charts without any trend. It may sometimes require the design of experiment in order to understand the root cause of such points.

STEP 9: IMPROVE THE PROCESS CAPABILITY ANALYSIS OF
        PROBLEM ROOT CAUSES.

Team work approach in all steps of implementing statistical process
control. However team work is crucial in this step if improvement
with each production cell is to be obtained. Each production cell
should be structured to include all members of both technical,
manufacturing and administrative support departments. Examples of
each production cells responsibilities are:

* Each production cell team to collect manufacturing inspection
  and machine operator data

* Calculate overall/high fliers process averages
  per sector to be plotted on control charts.

* Perform root cause analysis and look for:
  1) Assignable causes - for process average points that
     are above the U.C.L. or below the L.C.L.

  2) Sporadic causes - for process average points above the
     U.C.L. or below the L.C.L. that are cyclic in nature.

  3) Maintainable causes - for process average points that
     are stable within the upper and lower control limits.

* Develop action plans to address process induced defects.

* Monitor all improvements based on actions taken.

STEP 10: IMPLEMENTATION OF CORRECTIVE ACTION AND ON-GOING
         PROCESS CONTROL MONITORING

Statistical process control in a group technology environment
requires an on-going process that emphasizes quality at the
source of production. The prevention system for defects is
necessary. All key actions implemented within the process
must follow a means of de-emphasizing the detection / reaction
system. Corrective action implementation requires total team
involvement and commitment. There should be assigned tasks
with responsibility. The key elements in the on-going process
control management is shown in Figure 4.

A CASE STUDY RESULT

The implementation steps described earlier were applied in a
manufacturing plant that produces printed circuit boards to
customer orders. The entire production process from raw
materials to finished goods was divided into four group
technology cells. The manufacturing process methods were
revised to enable complete cell balance based on production
parameters.

254

For each cell, control charts and measures were
developed to monitor the process performance and yield.
Computer data base was developed for each cell. Within each
cell both process and system enhancement was made. The results
obtained after the implementation of statistical process control
in each cell is shown in figures 5, 6, 7, and 8 respectively.
In addition to the significant improvement in quality in each
group technology cell, it was observed that partial productivity
(labor productivity) in each cell increased significantly after
the implementation of statistical process control.
Other benefits obtained were: reduction in cycle time, and
inventory, improvement in percent of good conforming printed
circuit boards, less rework, and reduction in setup times. The
production planning process for the entire group technology
environment was also made easier after the implementation of
statistical process control. Statistical process control enabled
the clear understanding and prediction of the parameters and
variables in each production cell.

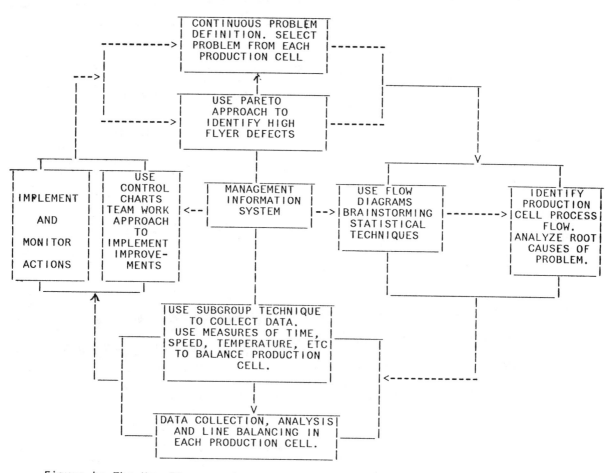

Figure 4: The Key Elements In On-going Process Control Management

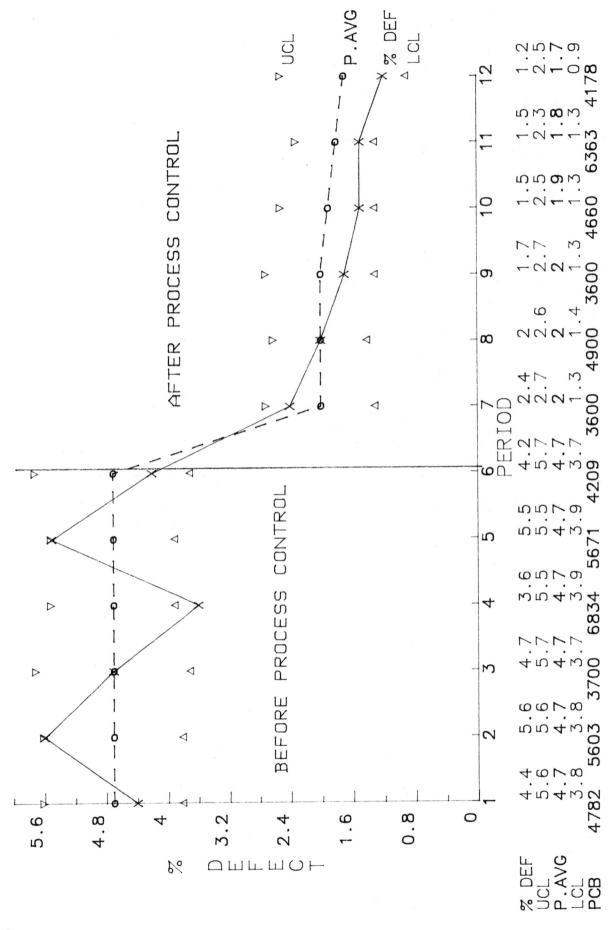

Figure 5: Statistical Process Control Result for GT Cell 1

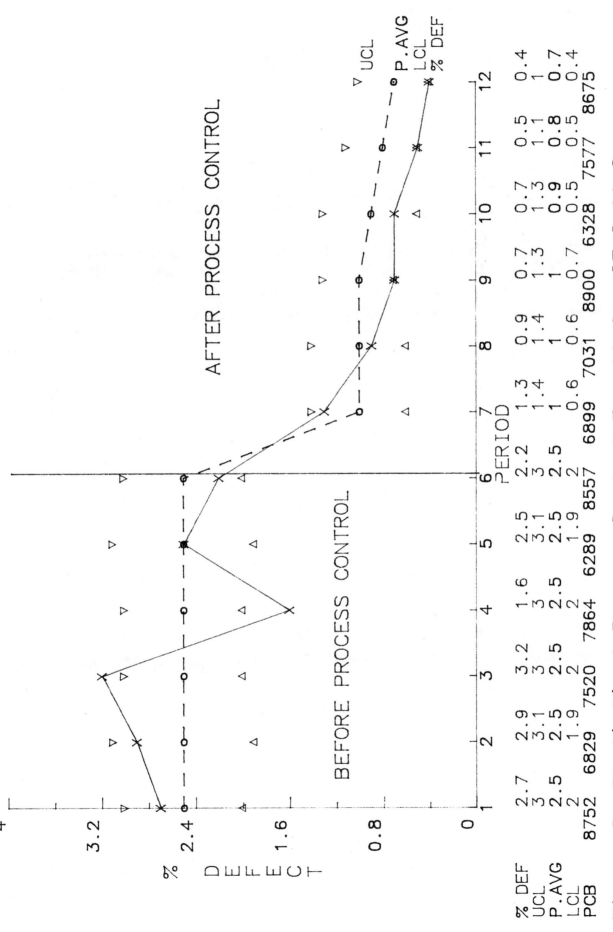

Figure 6: Statistical Process Control Result for GT Cell 2

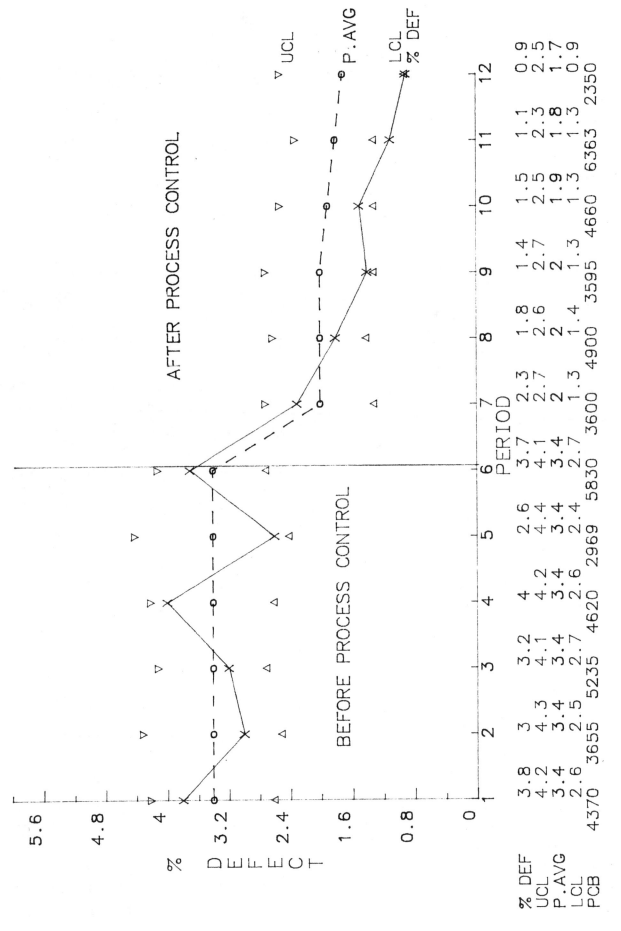

Figure 7: Statistical Process Control Result for GT Cell 3

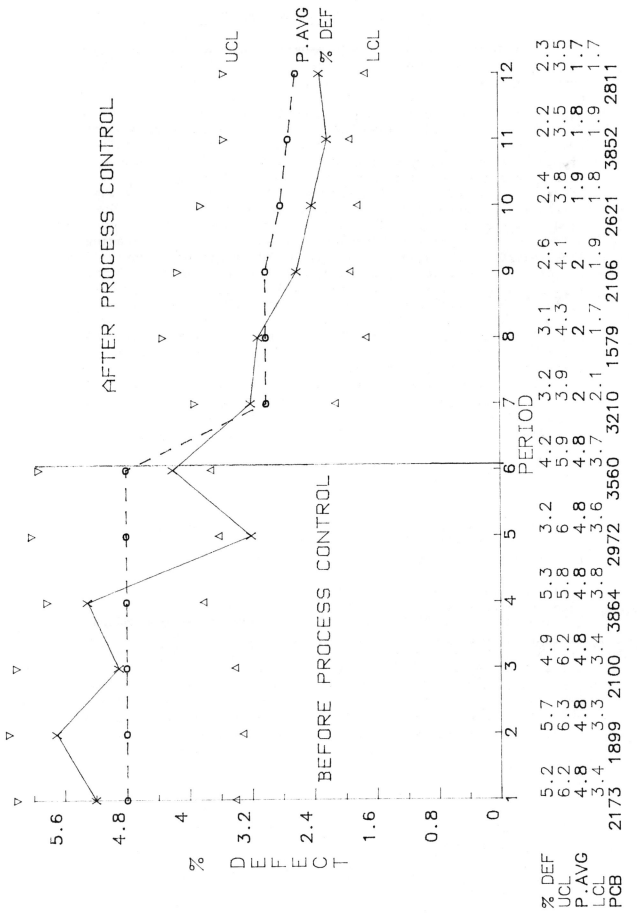

Figure 8: Statistical Process Control Result for GT Cell 4

PROBLEMS ENCOUNTERED DURING IMPLEMENTATION

The problems presented here are mainly intended to form a guide to companies implementing statistical process control in a group technology environment. It must be pointed out that problems associated with statistical process control in GT environment could vary depending on the organization setting and production environment. Based on the case study presented, we found that incoming product quality posed a significant problem at the machine level. This problem was resolved by characterization of products by vendors and front end screening before products are allowed in each GT cells. However, the vendors were finally made to take ownership for the correction of incoming defects. Grouping the processes, machines, products, etc. into GT cells could be a tremendous task. To overcome this problem, a team work approach was used in the cell balancing requirements. This also enabled any resistance to the study to fade away. Data collection, compilation and computation was also found to be cumbersome. A computer program was developed for this purpose which eased the data management burden. The characterization of each GT cell required a flexible work schedule driven by high customer demands for existing products. This problem was overcome through rotational work schedule among people and departments. Initially, crisis developed when a problem in a particular GT developed without clear understanding of the cause and who and what was responsible. To overcome this problem, specific tasks were assigned to specific people and meetings were held regularly to discuss and resolve problems real time. Training provided for both management and employees to facilitate good root cause analysis for problems and implementation of corrective actions.

CONCLUSIONS

Group technology (GT) approach to production management is a technique that will become increasingly important in the era of technological explosion. GT by itself does not offer greater advantage than other traditional approaches. However, with the implementation of statistical process control in GT environment tremendous benefits and potentials exist in productivity and quality improvement. The concepts of statistical process control in a GT environment is also still in its earlier stages of testing and implementation. More work is still needed in providing a system based approach for controlling and managing the parameters within each GT cell. This paper has offered an approach that has wide range of application in various GT groups; CAD/CAM, robotics, etc. The success in the use of the approach suggested depends among others on management and employees' commitment to excellence through quality and productivity improvement at the source of production.

BIBLIOGRAPHY

(1) Burbidge, J.L. (1975).The Introduction to Group Technology, John Wiley, New York.

(2) Edosomwan, J.A. (1985). "The Meaning and Management of Productivity," Industrial Engineering Working Paper.

(3) Edosomwan, J.A. (1985). "Quality At The Source of Production,"
IBM Technical Working Paper.

(4) Ham, I. (1976). Introduction to Group Technology," Technical
Report MMR76-03. Dearborn, Michigan: Society of Manufacturing
Engineers.

(5) Rodriquez, R. and Adaniya, O. (1985) "Group Technology Cell
Allocation," 1985 Annual International Industrial Engineering
Conference Proceedings.

(6) Deming, W.E. (1982) Quality, Productivity, and
Competitive Postition. MIT Press.

BIOGRAPHICAL SKETCH

Dr. Johnson Aimie Edosomwan is the Manager of the Manufacturing
Process Engineering, at IBM Data Systems Division in New York. He
received a Doctor of Science in Engineering Administration, with a
minor in Economics from the George Washington University, Washington,
D.C. He also received a P.Engr. degree in Industrial Engineering from
Columbia University in New York, M.S.I.E. and B.S.I.E. degrees, a
Certificate in Safety and Health, and a Certificate in Productivity from
the University of Miami, Florida. A.A. (Arch. Engr.) from Miami Dade
Community College, Florida; a Certificate in Management from AMA, and a
Certificate in Productivity and Quality Management from the George
Washington University. He is a graduate of the IBM Management School,
Armonk, New York; the IBM Robotics Institute, Florida, and the IBM
Quality Institute, Connecticut. He has worked as a consultant for
various consulting firms such as: Arthur Young and Company, Touche Ross
and Company, and Traintex Management Services. He has taught both at
the University of Miami and Polytechnic Institute of New York. He
is a recipient of several academic fellowships such as: Burger King
Fellowship, NIOSH Fellowship, IBM Fellowship, Social Science Research
Council Fellowship (U.S. Dept. of Labor). Selected awards and honors
are: IBM Engineering Quality Award, IBM Cost Effectiveness Award,
Outstanding Alpha Pi Mu Member of the Year Award, Special Award for
Contribution to the University of Miami (I.E. Department), and
Outstanding Service Award from the University of Miami Graduate School.
A member of Omicron Delta Kappa, Tau Beta Pi, Alpha Pi Mu, and Sigma Xi
honor societies. He is a senior member of IIE. He has been guest
speaker in several professional seminars and conferences worldwide and
has published several technical papers. His areas of interest are
productivity, quality and technology management.

Presented at the SME/FMA FABTECH '86 Conference, June 1986

Statistical Process Control (SPC) and Automation

Hans J. Bajaria
Multiface, Incorporated

Among many viable approaches to improve Quality and
Productivity of the manufacturing processes in automotive related
industries, the recent thrust continues to be implementation of
Statistical Process Control (SPC) and Process Automation. These
two items are consuming a significant portion of the automotive
industry investment funds. For such an investment to pay off in a
timely manner, the inherent ability of process automation to
provide a higher level of quality must be integrated with SPC.

In this paper we examine different schemes for integrating
SPC and automation. The various elements of the schemes considered
are: Forms of Data Collection, Types of Data Analysis and Plotting,
Types of Signals for Detecting Incipient Troubles, Discovery of
Associated Disturbing Causes and Correction of those Causes. For
this integration to succeed, an understanding of both functional
and process control automation and their balance is also needed.
We discuss the fundamental differences between these two forms of
automation. Further we elaborate on the distinction between the
deterministic and probabilistic process control automation and
their balancing criteria. To avoid a disproportionate distribution
of the invested funds it is necessary to optimize process performance
with respect to three forms of automation: functional automation;
deterministic process control automation; and probabilistic process
control automation.

This paper provides guidance to facilitate automation of
manufacturing processes or to improve the efficiency and yield of
existing automated processes.

INTRODUCTION

Automation has the potential to improve the productivity of
manufacturing processes and simultaneously to increase the quality
of their output. In order to tap the potential benefits of process
automation, various time consuming functional elements and various
causes that contribute to unacceptable process performance must be
understood and controlled.

Productivity improvement is based on the speed with which
automated process can operate as compared with manual processes.
When automation offers speed, it can be labeled as FUNCTIONAL
AUTOMATION. On the other hand, the quality potential of process
automation is based on the precision with which an automated
process can repeat as compared with human hands. Therefore, when
the automation is applied to improve the consistency of process
output, it can be labeled as PROCESS CONTROL AUTOMATION. Figure 1
depicts the difference between FUNCTIONAL AUTOMATION and PROCESS

CONTROL AUTOMATION.

FUNCTIONAL AUTOMATION

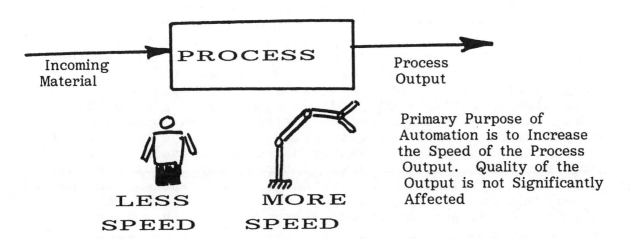

Primary Purpose of Automation is to Increase the Speed of the Process Output. Quality of the Output is not Significantly Affected

PROCESS CONTROL AUTOMATION

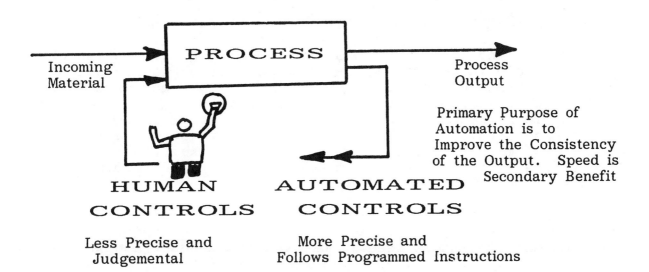

Primary Purpose of Automation is to Improve the Consistency of the Output. Speed is Secondary Benefit

Figure 1 - Difference Between Functional Automation and
Process Control Automation

The primary focus of functional automation is to increase the speed of the process whereas the primary focus of process control automation is to improve the consistency of process output. Regardless of where the focus is intended to be directed, process automation always results in significantly greater speed than can be achieved by human hands. In this paper, the emphasis is placed on process control automation.

It is also important to understand the role of Statistical Process Control (SPC) as it relates to process control automation. Use of SPC to monitor any process output will indicate the instabilities as well as out-of-control conditions in the order that they occur. As the causes for such conditions are discovered, corresponding corrective actions need to be found and taken to restore the process to its natural state. If such conditions occur on a consistent basis, there must be standard operating procedures developed to take care of them. These procedures then can be automated so that their execution will not depend on human judgement or motivation. This scenario forms the close connection between automated process controls and SPC. Statistical control charts can be considered to provide the strategic guidance necessary to convert those standard operating procedures into automated procedures in the order that they create disturbances to the process output. The use of automation for process control generally requires greater capital investment than conventional ways of handling the same issues. If the investment is required to yield a timely payoff, it must address those hardware improvements that bring quicker rewards. The use of SPC in determining what aspects of process controls need to be automated first, almost guarantees a handsome return on investment.

Basically process control issues can be broken down in two categories: (1) Process control against previously known process disturbances and (2) Process control against previously unknown process disturbances.

(1) PROCESS CONTROL AGAINST PREVIOUSLY KNOWN DISTURBANCES

Figure 2 shows the three elements of process control which can play a role in perturbing the condition of a process output, namely, (A) Human Hands, (B) Standard Operating Procedures (SOPs), and (C) Process Parameters. It is necessary to control the known relationship between each one of the process control elements and the process output. That means, if efforts are not made to maintain these relationships, the process is likely to loose its state of statistical control. These are known as Deterministic Controls.

(A) Role of Human Hands - It is a well-established fact that automated equipment will repeat better than human hands. Suppose that a manual process is used to produce the product and SPC is used to track its output. With sincere use of SPC by operating personnel, it would be possible to keep the process in a state of statistical control. However, if one examines the manual process capability, it will certainly be wider than that which can be achieved by the use of mechanization or robotization. Therefore, if automation is used instead of human hands, not only can the process be maintained in a state of statistical control but its capability will be much improved (i.e. 6 sigma will be narrower).

There are two points to remember as to how automating the role of human hands and SPC are tied together.

a. SPC can be used to ascertain whether the process is going out of control as a result of variation in manual handling or that the process has an unacceptably wide spread. Automating the process can solve both problems and improve the process significantly. Here SPC can act as a thermometer indicating type and degree of sickness, whereas automation can act as medication for removing that sickness.

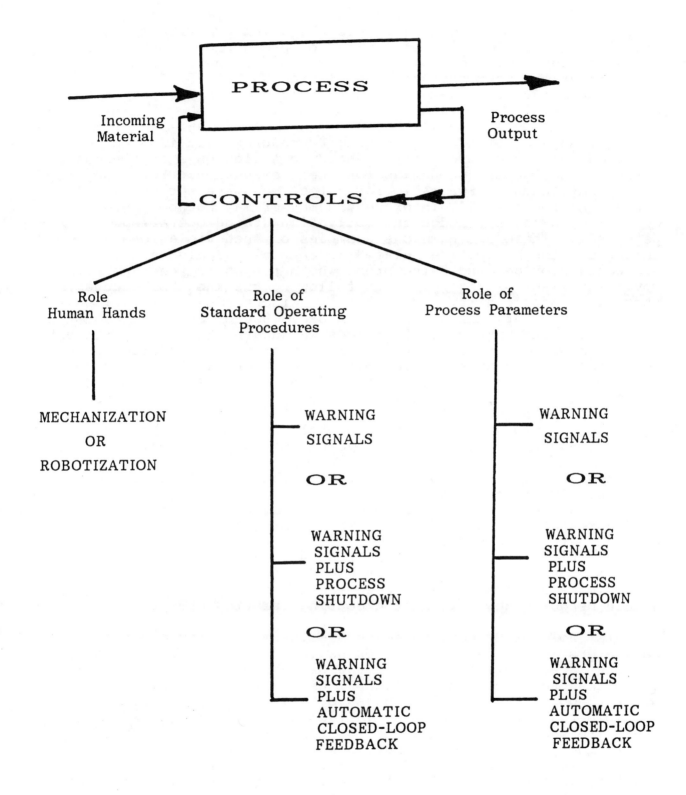

Figure 2 - Deterministic Process Controls and Automation

b. If there are some products which a competitor makes using automated machinery, he clearly has the advantage with respect to process repeatability and efficiency of production. SPC of the manual process will not compete with such a scenario eventhough it can deliver the best that the manual process has to offer. The use of SPC cannot be considered to be strategic under this set of circumstances.

(B) Role of Standard Operating Procedures - It is a well-accepted fact that SOPs are very hard to follow on a continuous basis if they depend on humans for their execution. For example, let us say that the SOP calls for a tool change after every 300 parts are processed. This procedure may or may not be executed depending on the knowledge and motivation of the operating personnel. There are numerous examples of such violations throughout the automotive related industries. Implementation of SPC would provide conclusive proof whether such is the case or not. Obviously, lack of discipline in following the known procedures results in poor quality. Using SPC to discover such violations is certainly not very effective use of SPC. Use of SPC will assist us to discover the unknown disturbances which are otherwise difficult to determine and not to prove the fact that the procedures are not followed. If the SOPs are automated then there would not be any need for human judgement in executing the SOPs. The automated SOPs can provide signals or signals plus automatic halt in the process or signals plus automatic correction of the disturbing causes. Of course, the automation is not completely foolproof either, but several orders of magnitude improvement can be expected.

Again, there are two points to remember as to how automating SOPs and SPC are tied together.
a. SPC will only reveal whether standard operating procedures are followed or not. The automation of SOPs will assure that they are followed.
b. Discovering violations of SOPs by using SPC is not a very effective use of that tool. Why use SPC to discover more things to do (more SOPs) when we know we are not doing the things we should be doing now (i.e. not executing existing SOPs).

(C) Role of Process parameters - There are many process parameters (e.g., temperature, pressure, feed, speed, cure time, loading rate, etc.) that affect the process output. These parameters need to be controlled within well-defined ranges in order to obtain acceptable process output. It is not always possible to keep these process variables within prescribed ranges. There are basically two reasons for this difficulty: equipment limitations; and over-eager but unskilled problem solvers who over-control the process. Use of SPC can point which case applies. However, if the process is automated, we can obtain a signal when such violation occurs or obtain a signal and shut down the process; or we can obtain closed-loop feedback to fix the disturbing signals.

Another important point to note which ties the automating process parameters ranges and SPC is that the use of SPC can reveal violations to prescribed process parameter ranges whereas automation can provide real time signals and closed-loop feed back for their correction when they are violated.

(2) PROCESS CONTROL AGAINST PREVIOUSLY UNKNOWN DISTURBANCES

Manufacturing operations are always prone to problems. Furthermore, the causes of these problems are often difficult to predict in advance. Causes which can be predicted are generally handled by Standard Operating Procedures, Prescribed Process Parameters Ranges, and Well-defined Training Elements for operating personnel. As discussed earlier, use of SPC will reveal whether the discipline exist to control these known relationships and the use of automation will almost guarantee that any disturbances to these relationships which can put the process statistically out-of-control will be better controlled. What about unknown causes that can disturb the state of control for any process? SPC is the only tool capable of providing true signals against disturbances of unknown origin. The execution of SPC requires that the process output (product characteristics) signals be monitored authentically and on real time basis on appropriate statistical control charts and that actions be taken based on the interpretation of these charts.

In the past, automotive companies have utilized control charts only on a selective basis rather than a rule. In fact, their focus has been on product specifications rather than statistical control limits. Corrective actions are taken only when the parts are out-of-specification range. Use of SPC requires that actions be taken when the parts are out of statistical control limits. Similarities between two approaches are that in both instances the part conditions are examined but the differences come in as to when the actions are initiated. This fundamental difference is very difficult to rectify in the automotive production environment.

A major difficulty in implementing SPC is that the workforce in automotive related companies is habituated to think specification lines and not control lines. Though the control chart concept can be illustrated with simple mathematics in the classroom, it is very difficult to execute the idea on the production floor due to "conventional thinking habits". Furthermore, decisions made using control lines, in some instances, coincide those that may be made using specification lines (or by the good old instincts of the experienced operators). This creates argument in favor of status quo. One can state that decisions based on control limits will always be correct, whereas decisions based on specifications limits will sometimes be correct and sometimes be incorrect. Regardless, if one were to use control limits as the basis for decision making for process corrective actions against unknown causes, then several steps need to be understood for proper execution of SPC. These are:

(a) Measure process output (product characteristics of interest)
(b) Record output as raw data
(c) Plot authentic summary of these data on the appropriate control charts on a real time basis
(d) Interpret control charts as each additional point is added to the control charts. For complicated patterns showing up on the

control charts, this may take considerable analysis time.
(e) Determine corrective action.
(f) Take corrective action.

In present day automotive production environment, if operating personnel were to execute SPC, they must be educated in SPC philosophy, and SPC methods. In addition, they must be trained in the execution of these methods in their production environment. This is a very sore issue not only for the automotive related companies but for any industry, since the majority of workforce coming from a diverse background is neither educated nor trained to think like this. Additionally, an organization issue that may hamper the SPC effort, is performance measurement by time standards. The makeup of standard time, in general, does not account for the time it may take to execute SPC steps effectively. Of course, there are some exceptions where an operator has idle time between operations. SPC efforts can be "squeezed into" that idle time. This may or may not be welcomed by the operator who may perceive this as merely an additional burden. Also, very few, if any organizations who claim to have implemented SPC have modified their standards. If one critically examines SPC implementation success claims made by several companies, there are bound to be some cases of exaggerations, since effective SPC implementation is usually not possible without modification of some of the time standards. Use of process control automation may provide some resolution for these difficulties.

Figure 3 indicates several stages of automation that are possible. This is also known as Probabilistic Controls in the sense that process output signals are statistically monitored and their interpretation (to determine the disturbing causes) requires the use of statistical tools.

Stage a - Process output is measured manually (i.e., the operator makes the measurement). The measurement device can be directly plugged into the data storing device which can then be used to generate appropriate control charts. This stage has the disadvantage that the operator still has to handle the measuring device.

Stage b - In this stage, the gage is part of the process. Process output is automatically measured and the desired statistical signal is stored in the data storing device. This data storing device itself can be capable of generating an appropriate chart or it can be plugged into a computer to do the same. This can cut down several manual steps that might be required in measuring, summarizing and plotting. Interpretation of the pattern is still based the skill of the chart reader, with the exception that an out-of-control condition can be signaled.

The disadvantage of this method is that the cost of gage (which is now an inherent part of the process) may be prohibitive. Also, it may not be possible to buy off-the-shelf measurement technology to measure all of the process output characteristics of interest.

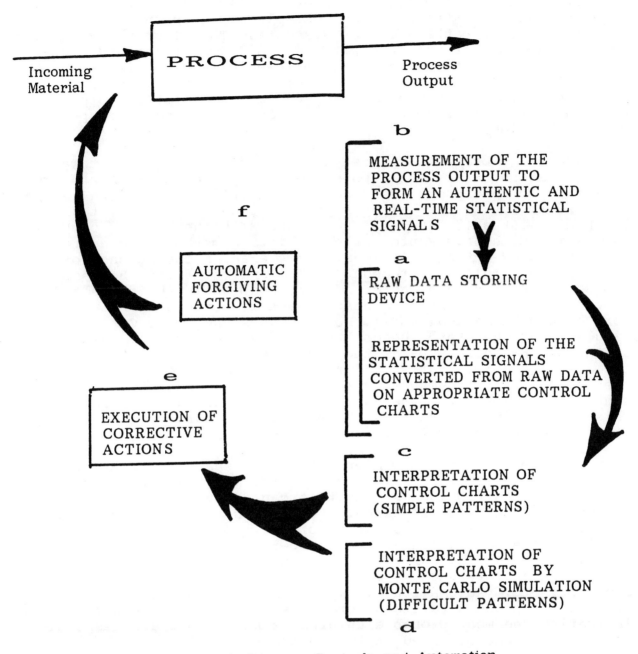

Figure 3 - Probabilistic Process Controls and Automation

Stage c - The question of interpretation has always bothered SPC enthusiastists. In fact, insurmountable difficulties in reading the charts and determining the corresponding hardware actions has resulted in apathy toward implementing SPC. What is required to interpret the charts is to be able to sincerely jot down the process changes as soon as they occur as well as track the condition of some top suspect variables along with each plotted point. Many failures in reading charts are associated with the incompleteness of such records. Here automation can help out again. One can build the idea of tracking potential suspects along with plotted signals on the control charts. Also one can build the algorithm into the computer to automatically signal the condition when the varying pattern on the control charts begins to coincide with the varying pattern of the suspects. The algorithm should be able to examine not only the patterns followed by the individual suspects but also those that are followed by all possible combinations of suspects. This automation in control chart interpretation can reduce the tedious task of relating output patterns with the suspect patterns to determine the corresponding hardware actions.

Stage d - Even after these stage c efforts are made toward interpretation, it is still possible that the control charts are difficult to interpret. Here control chart interpretation can be advanced to the next level of sophistication. The analyst should be able to ask some "what if" questions with the help of computer simulation methods. Monte Carlo simulation methods are quite powerful in executing this idea. Lets examine the essence of the Monte Carlo simulation procedure. The analyst first chooses (guesses) the type and degree of perturbation in the suspected process variables. Then he describes this perturbation in the form of a mathematical model. He then uses the Monte Carlo simulation to generate the control charts that reflect this scenario. He then compares the simulated control charts with the actual control charts. If the simulation generated based on one or more of his guesses in fact duplicates the actual (observed) control chart pattern, the analyst can then explain the observations. This is the ultimate that can be achieved with respect to automation of control chart interpretation.

Stage e - One can now step into the arena of hardware action to correct the process. After having interpreted the control charts, SPC has provided all that it can deliver. Corrective action still needs to be taken. Here again automation can help. Process corrective actions can be of two types: turnable knobs to alter process parameters such as temperature, pressure, cure time, speed, feed, loading rate, etc; and unturnable knobs to alter process parameters such as material variation, worn tool, fatigued operator, etc. Automation can maneuver the turnable knobs to restore the process to its natural state or it can flash a message indicating the most probable unturnable knobs that are responsible for the patterns observed on the control charts.

Stage f - The ultimate in automation is to be able to utilize the forgiving nature of some process variables. For example, if harder material than usual needs to be processed, this condition can be detected by the integral process control charts and correspondingly the speed of the machine can be reduced to get the acceptable process output signals. Of course, in order to achieve such counter actions by the forgiving process parameters (correction of process speed to compensate for the hardness of the material), both process knowledge and process flexibility must exist. For many manufactured products such process knowledge either already exists or can be developed with the use of statistical methods. The issue of process flexibility is not very difficult.

270

Predicting the unknown has never been easy, but statistical thinking (SPC) and high speed computers (automation) at least make the task approachable for most of the unknown problems, and solvable for some problems which cannot be approached by any other equally effective means.

SUMMARY

Companies in the the automotive industry are spending lots of money trying to improve the productivity with which cars are placed on the market. Among many productivity ideas being implemented throughout this industry, the Statistical Process Control and Automation are consuming a significant proportion of the budget.

This paper has analyzed the close relation between Automation and SPC. SPC provides the guidance as to what needs to be done whereas automation makes sure that it gets done!

This paper also described the distinction between Functional Automation and Process Control Automation. In functional automation the speed or efficiency of the process in generating output is a primary focus and the output consistency is of little or no consequence. In Process Control Automation, the consistency of the output is a primary focus, whereas the increased speed of the process output is a secondary benefit.

Process Control automation is further broken down into Deterministic Controls and Probabilistic Controls. Deterministic Controls are those that control the well-established relationships between causes and process output. If these relationships are disturbed, the statistical signals can be picked up and either an automatic feedback loop can be provided to make a necessary process correction or the process can shut down to permit a manual process correction. Use of SPC in such cases, can only establish the priority with which the funds should be invested in automating deterministic controls. Probabilistic Controls, on the other hand, are those that first requires us to make an attempt to understand the relationships between the process output and the disturbing causes through the use of control charts. Several steps are required in executing SPC on a given process, starting from measuring process output to determining and correcting the disturbing causes. These steps can be automated by installing probabilistic controls. Several stages of Probabilistic Control Automation were discussed. There is no effective way, other than SPC, to achieve this.

COMMENTARY

Though automotive related industries are applying good ideas in form of SPC and AUTOMATION to improve productivity, they are not necessarily doing it in correct strategic order. As a result, there is a need to question and further understand the relationship between SPC and AUTOMATION.

First of all, there is plenty of evidence to indicate that

the robotization of a manufacturing processes is usually implemented with the primary focus on the speed rather than the consistency of process output. One cannot and should not argue against the benefits of speedy operation, if a proportionate amount is also invested in PROCESS CONTROL AUTOMATION. Such is not the case, however. There is sufficient evidence to suggest that a disproportionate amount of funds are being spent on Functional Automation as compared with the spending on Process Control Automation.

Secondly, there is also evidence that the funds going toward Process Control Automation are disproportionately distributed in favor of Deterministic Process Controls. Actually, there is an acute need to invest in understanding the relationships between the patterns in the process output and the corresponding contributing causes rather than to invest in deterministic controls. This can be achieved by the use of probabilistic controls. If the process of applying probabilistic controls is speeded up through the use of automation, there are more benefits to be derived than to simply install deterministic controls without knowing the priorities. The priorities for installing deterministic controls can only be established through the use of probabilistic methods.

Therefore, any company considering automation must first understand the balance that must exist between investing in Functional Automation, Automated Deterministic Process Controls and Automated Probabilistic Process Controls. If it is unclear as to how funds should be distributed among these three forms of automation, SPC should serve as a conclusive guide toward establishing priorities.

Automotive related industries know that their survival is at stake. They know that the funds need to be invested. But they need to be invested strategically in upgrading the machinery that produces products. They also know that this machinery must be automated to compensate for the high cost of labor and the inconsistency of output compared with overseas competition. What they don't quite understand is the delicate investment balance that must exist in implementing various phases of automation. To understand the power that lies in talking about SPC and automation together, they should focus ON DOING THINGS RIGHT RATHER THAN DOING THINGS FAST.

If additional information is requested please feel free to contact us.

Fabricators and Manufacturers Association, Intl'
5411 E. State Street
Rockford, IL 61108
815/399-8700

INDEX

engineering, 63
fatigues, 64
in-process, 62, 64, 69
mold, 122
packaging, 42
part, 122
production workers, 20
prototype, 42
receiving, 68
stations, 113
supervisors, 42
Internal newspapers, 6
Inventory, 68

J

Japan, 19
JIT, See: Just-in-time
Joining, 68
Just-in-time, 5, 229

K

Keyboards, 217

L

Labeling, 39
Labor, 23, 25, 229, 245
Laminate fabrication, 123
LAN, See: Local area networks
Lasers, 70-71, 230
Local area networks, 230
Loss function approach, 34
Lot tolerance percent defective, 106
Lubricants, 176

M

Machinery, 76
Machining, 122
Maintenance, 7, 42, 57
Management
    attitude, 8
    authoritative, 21
    commitment, 49
    contract, 5
    conventional, 31
    failure, 31-32
    improvement, 33
    issues, 27
    levels of, 9
    and motivation, 9, 11
    operating, 31
    organization, 21
    participative, 21
    philosophies, 21
    problem solving, 11

process, 247
production, 260
reports, 41
responsibility, 175
short-term, 4
standard, 31
tactical staff, 36
top, 5, 16, 56, 179
total, 32
of "unquality", 42
Management information systems, 255
Manual SQC methods, 57
Manufacturing automation protocol, 70, 230
Manufacturing cells, 250, 253
Manufacturing cycle, 17
Manufacturing processes, 5
Manufacturing systems, 62-72
MAP, See: Manufacturing automationn protocol
Marketing, 5, 9
Material curing, 122
Material receiving, 122
Material review, 68
Material status alteration, 68
Measurement, 163
Mechanical finishing, 180
Mechanical tests, 136
Media, 185, 189
Media mobility, 185
Metal finishers, 162-171
Metal removal, 68
Metrology, 11, 42, 45
Micrometers, 53
Microprocessor technology, 171
Microscopic examination, 124
Minicomputers, 57
Mixed models, 98
Mold design, 122
Molding, 139
Molding process, 124

N

NC, See: Numerical control
Nested designs, 102-103
Nondestructive testing, 122
Nonsynchronous power, 233
Normal distribution, 92
Normal inference chart, 60-61
Nozzles, 96, 99-100
Numerical control, 58
Numerical Quality Index of Customer Performance, 10

O

Ongoing process control, 86
Operations support, 5
Operator interface, 213
Outputs, 59

P

Packaging, 39, 42
Packing, 68
Paint, 162-171, 208
Panel protocol, 155
Parameter designs, 34
Part fabrication, 122
PCB, See: Printed circuit boards
Peak temperatures, 168
Percentage defectives, 115
Performance measurement systems, 30
Philosphy, 56
Pick and place, 232-233
Picking, 68
Planning, 8, 17
Plastics, 68
Plotters, 237
Polar deviation, 60
Policy control, 13
Policy development, 13
Polling systems, 216
Polyester/glass laminates, 125-127
Powder compound, 188
PREP, See: Product Review and Evaluation Program Report
Prepreg material, 123
Preproduction planning cycle, 17
Pressure, 235
Preventative maintenance, 42, 57
Printed circuit boards, 245-261
Printers, 237-238
Probabilistic controls, 268
Probability plots, 59-60
Problem prevention, 28
Problem-solving, 19
Process average, 90
Process capability, 57-58, 93
Process capability studies, 57-58
Process center lines, 83
Process control, 48-56, 86, 172-179, 212-228, 245-262, 267
Process control automation, 262
Process parameters, 248
Process variation, 75
Product acceptance, 42
Product assurance, 5
Product design, 7
Production, 7
Production operations, 5
Production process, 51
Product liability, 245
Product reliability, 21
Product Review and Evaluation Program Report, 10, 38
Product test data, 3
Profits, 16
Project control, 29
Proportion defectives, 88-89
Protocols
    cycle time, 156
    humidity cabinet, 155
    manufacturing automation, 70, 230
    panels, 155
    run, 156
    salt solution immersion, 156
    scab corrosion test, 155
Prototype inspection, 42
Prototypes, 39

Q

Quality control circles, 19
Quality function deployment, 12

R

R & D, See: Research and development
Random angles, 100
Raw materials, 5, 76
Real time, 9, 57-58
Rejection, 10, 48, 52
Rejection reports, 10
Reject rates, 48
Reliability analysis, 3
Reliability assurance function, 18
Reports
    customer complaint, 26, 36
    economics of quality, 26
    product review, 10, 38
    shop floor, 11
    systems, 11
Research and development, 9
Resin content, 134
Return on investment, 231
Return-to-vendor, 69
Robots, 8, 66-67, 70, 212-218, 229
Run protocol, 156

S

Safety, 122
Salt solution immersion protocol, 156
Salt spray tests, 143-161
Samples, 122
Sampling, 48-56, 106, 172-179
Sampling frequency and the sample size, 82
Sampling plans, 108-109
Scab corrosion test protocol, 155
Scrap, 45, 246
Sealing machine, 102-103
Sensing systems, 70
Sensors, 205, 208, 210
Setup operations, 58
Setup time, 171
Shipment, 7
Shipping, 68
Shop floor quality reporting system, 11
Short beam shear, 127
Short-term financial incentives, 14
Short-term goals, 14
Small parts, 69
Software, 60, 167-168, 204-205, 212-229, 237-240, 242
Sources of variation, 75
SPC, See: Statistical process control
Specimen geometry, 135
Speed, 181
SQC, See: Statistical quality control
Standard deviation, 94
Standard normal distribution, 92
Standard operating procedures, 266
Standards, 22

T

V

W

Z